Contesting Leviathan

Contesting Leviathan

Activists, Hunters, and State Power in the
Makah Whaling Conflict

LES BELDO

The University of Chicago Press
Chicago and London

PUBLICATON OF THIS BOOK HAS BEEN AIDED BY A GRANT FROM THE BEVINGTON FUND

The University of Chicago Press, Chicago 60637
The University of Chicago Press, Ltd., London
© 2019 by Les Beldo
All rights reserved. No part of this book may be used or reproduced in any manner whatsoever without written permission, except in the case of brief quotations in critical articles and reviews. For more information, contact the University of Chicago Press, 1427 E. 60th St., Chicago, IL 60637.
Published 2019
Printed in the United States of America

28 27 26 25 24 23 22 21 20 19   1 2 3 4 5

ISBN-13: 978-0-226-65737-0 (cloth)
ISBN-13: 978-0-226-65740-0 (paper)
ISBN-13: 978-0-226-65754-7 (e-book)
DOI: https://doi.org/10.7208/chicago/9780226657547.001.0001

Library of Congress Cataloging-in-Publication Data

Names: Beldo, Les, author.
Title: Contesting Leviathan : activists, hunters, and state power in the Makah whaling conflict / Les Beldo.
Description: Chicago : The University of Chicago Press, 2019. | Includes bibliographical references and index.
Identifiers: LCCN 2019009777 | ISBN 9780226657370 (cloth : alk. paper) | ISBN 9780226657400 (pbk. : alk. paper) | ISBN 9780226657547 (e-book)
Subjects: LCSH: Whaling—Washington (State)—Makah Indian Reservation. | Makah Indians—Fishing. | Whaling—Moral and ethical aspects—United States. | Whales—Conservation—United States. | Whaling—Government policy—Washington (State)
Classification: LCC SH383.2 .B45 2019 | DDC 639.2/809797—dc23
LC record available at https://lccn.loc.gov/2019009777

♾ This paper meets the requirements of ANSI/NISO Z39.48–1992 (Permanence of Paper).

CONTENTS

INTRODUCTION / 1

ONE / It's Who We Are / 32

TWO / We Eat Them / 61

THREE / Everything Is Connected / 85

FOUR / This Fishery Will Be Managed / 107

FIVE / You Just Don't Kill Whales / 121

SIX / The Science Has Ruled / 137

SEVEN / The Whale Approaches / 152

CONCLUSION / 171

Acknowledgments / 187
Notes / 191
Bibliography / 215
Index / 227

INTRODUCTION

On May 17, 1999, a whaling crew from the Makah Indian Nation hunted and killed a gray whale off the coast of Washington State. It was the first whale hunt by the Makah, or by any Native American tribe on the US mainland, in over seventy years. News helicopters circled overhead, allowing television viewers across the Pacific Northwest to watch live as the whalers harpooned the whale from a dugout canoe before shooting the whale with a .577-caliber rifle from a motorized support boat. For the previous two days of the hunt, the whalers had chased in vain as a group of gray whales fed in the shallow ocean waters off Cape Alava, near the Makah Reservation. A whale would surface, a plume of mud would blossom around it as it gulped down a huge mouthful of tiny crustaceans, and then the whale would disappear beneath the waves before the whalers, paddling furiously, could reach it. This happened again and again until, on the third full day and perhaps the thirtieth hour of the hunt, a juvenile female gray whale approached the canoe and lingered near the surface for several minutes. The whalers waited for the mandatory US federal observer to arrive before harpooning the whale.

In a community steeped in songs and stories of whales and other animals willingly sacrificing themselves to worthy hunters, the whale's behavior was easily explained. "Did you see," a Makah man asked me a decade later, "on the video, the way that whale turned, lifted its fin, and *received* the harpoon? Now that's [an] answer to prayer."[1]

Lawsuits filed by antiwhaling activists shut down Makah whaling the following year. The next Makah hunt did not take place until 2007, when five Makah whalers killed a gray whale in the Strait of Juan de Fuca, this time without the official sanction of the US federal government or the formal approval of the Makah Tribe. According to witnesses, that whale also swam up to the boat before being harpooned: "It came right to us," one of the harpooners later told me. "It chose us," said another to a reporter just hours after the hunt.

Local antiwhaling activists advanced a different interpretation of the whale's behavior. "That gray whale *calf*," one activist said of the 1999 hunt, emphasizing the whale's size and age (thirty feet and about two years, biologists estimated), "just wanted to be petted, not stabbed and shot with an elephant gun." The young whale's gregariousness reminded some activists of their visits to the birthing lagoons at the southern extent of the gray whale's range, where female whales often approach boats of awed tourists and nudge their calves up out of the water as if to be touched and petted. The behavior has earned these particular gray whales a nickname: the friendlies. Some activists speculated that the whale killed in the first Makah hunt might have been one of the friendlies, only recently petted by ecstatic tourists in its first few months of life.

"They say they understand the whale, that nobody knows the whale better than they do," another activist said to me of Makah whalers as tears welled up in her eyes. "But I know what it wants. It wants to live and eat and be with its family. You don't tell me it wants to be stabbed and blown up!"

Anthropologists more or less agree that *nature*, at least as we tend to understand and manipulate it in our daily lives, is not something that exists independent of human involvement.[2] The distinction between nature and culture is, itself, cultural.[3] Doing away with this binary has proven challenging, however, because it means we can no longer appeal to a realm of incontestable facts in order to settle disputes about things in the world.[4] What we see in nature reflects "all kinds of prior commitments to ways of seeing, studying, and classifying life" (Jasanoff 2012, 160). Classification tells us as much about the classifiers as the classified (Ritvo 1997, xii). Scholars have used a variety of metaphors to make this point, describing nature as "constructed," "performed," or "enacted."[5]

The fact that nature is socially constructed does not mean that it is purely a figment of our collective imagination. Surely, the two-year-old gray whale that approached the Makah canoe in the 1999 hunt—for the sake of convenience, I'll call her Yabis, the name given to her posthumously by a Makah elder who opposed the hunt—preexisted her encounter with Makah whalers.[6] Before she approached the canoe—before she was photographed, harpooned, towed to the beach, photographed some more, sliced, and divided, her tissues dissected, tested, and consumed—she existed, a singular living being at the center of her own world. As William Cronon writes, speaking of rocks, trees, and birds, but also presumably of whales:

> They are first and foremost themselves, despite the many meanings we discover in them. We may move them around and impose our designs upon them. We may do our best to make them bend to our wills. But in the end they remain inscrutable, artifacts of a world we did not make whose meaning for themselves we can never finally know. (1996, 55)

Although Yabis swam in acidified water, had carried deposits of heavy metals in her fatty tissues since before her birth, and subsisted in a food web irrevocably altered by commercial fisheries, it is nonetheless undeniably true that some part of her remained as Cronon's last line suggests: inscrutable, a singular living being, an existence that refuses to be conceptualized.[7]

It was only when Yabis approached the canoe, however, that she became so many of the things we think of when we think of gray whales. It was only then that she became *a whale*, and a North Pacific gray whale, and (very likely) a member of the Pacific Coast Feeding Group—the contested abstraction that has been the focus of so many legal debates over Makah whaling since 2002. She also became meat—food, sustenance for a community—just as for others she had only recently become the concrete subject of a particular kind of moral concern. In fact, only after she approached the canoe did the whale become Yabis.[8] This moment of many *becomings* frustrates what Bruno Latour would call the Moderns' desire to keep science, culture, and politics conceptually distinct. These distinctions blurred as Yabis became a character in cultural and historical narratives, an instantiation of a spiritual being, and also, briefly, a player in a hotly contested political and moral dispute. One way to take these different enactments seriously is to view them as ontological rather than epistemic—that is, to regard the many manifestations of Yabis not as different perspectives on the same passive object but as constituents of a reality that is multiple (see Mol 2002).[9]

Multiplicity is not something that any actual party to an ontological dispute is obliged to accept, however. In fact, if Yabis had a precursor, it was another whale-in-the-singular: it was *the whale* that Makah hunters and antiwhaling activists had been talking about for years before the 1999 hunt. So many of the back-and-forth exchanges between Makah whalers and antiwhaling activists in person and in print invoke this figure of "the whale" in the singular, as if there were only one, or as if the living being that would end up at the other end of the harpoon had already been identified and selected.

"Just remember," a Makah spokesperson told a visiting journalist almost two years before the first hunt, "the whale knows it's being hunted."[10]

"I know what the whale wants," goes the activist reply.

Such claims provide ontological definition. To know what the whale wants is to lay claim to its real essence—to deny, sometimes expressly, competing enactments of it. Such claims do not suggest a desire to participate in what Mol calls ontological politics (a politics of what *is*) as much as a desire to bring ontological politics to an abrupt end by defining the whale's one true enactment.

It gets more complicated. Multiplicity may be a feature of ontology in complex societies, but in the Makah whaling conflict, *number* is also part of the stakes. Whether gray whales deserve moral consideration apart from their conservation status—that is, whether individual whales should be protected from harm, regardless of their numbers—is the kind of question that marks the fault line between the so-called conservationist and preservationist positions in environmental ethics. In this case, it is the question that separates Makah whalers and the US federal government from antiwhaling activists.

"We're not taking the last one!" Keith Johnson, the president of the Makah Whaling Commission, said to me on numerous occasions whenever the topic of opposition to Makah whaling came up. Other Makahs I spoke with simply could not understand how antiwhaling activists could seriously oppose a hunt on a species that is not legally classified as endangered or threatened. At the same time, antiwhaling activists express their frustration over a federal government that never seems to care about a whale being killed, just *how many* whales. Hence the activist retort, "One whale is too many," a qualitative argument in quantitative form.

It is fitting that the mandala in figure 0.1 depicts a single whale rendered in the "spy-hop" position—a slow, vertical surfacing behavior that cetologists believe to be a sign of curiosity—at the center of an image in which nearly every other actor appears in the plural. Painted by an activist leader after the 1999 hunt and not originally intended for public consumption, the exhaustively detailed rendering might be viewed first as an aspiration. For the whale to be seen in the singular and not solely as a quantifiable natural resource or fungible token fairly accurately sums up the original goals of antiwhaling activists in the Makah whaling conflict.[11]

Unfortunately for the activists and fortunately, I will argue, for Makah whalers, seeing whales in the singular is not something the state is inclined to do; in fact, it is predisposed against this very thing. Of the different ontologies that the whale approached when she swam toward the Makah canoe,

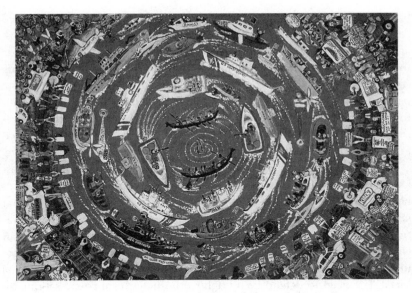

Figure 0.1 Margaret Owens, *Mandala* (2000)
Image courtesy of the artist

the US federal government's enactment of gray whales places the strictest limitation on her biographical possibilities as a moral subject. Whales have served as charismatic icons of the global environmental movement since its emergence, the imperative to "save them" a metonymic rallying cry for the Earth in its entirety, but the US federal government's approach to managing and caring for gray whales exhibits the same quantifying impulse that scholars have documented in myriad other projects of the modern state, from forestry to farming to fisheries.[12] The state sees gray whales as it sees most other aspects of nature: as resources, in accordance with the impulse to, as the eco-critic Rob Nixon (2011) puts it, abstract in order to extract. Whales have been called "near-sacrosanct" recipients of empathy and protection (Blok 2007, 66), and this may be true on Greenpeace posters and in BBC specials. For the US federal government, however, whales are countable, killable units arranged into "stocks," their fates discussed in a technical register saturated with agronomic metaphors like "harvest" and "yield." In short, for the US federal government, whales are still essentially fish.[13]

How whales have remained fish (i.e., marine resources), morally speaking, within what we might call the moral economy of the state would make for a compelling case study in the momentum (or perhaps the creeping expansion) of capitalist-cum-bureaucratic rationality, since whaling has not been commercially important in the United States for at least one hundred years. This book is less concerned with the reasons for this enduring discrepancy between the state's ontology and much of the general public's attitude toward whales, however, than in its sweeping effects on the Makah whaling conflict and others like it—conflicts involving animals and the state where parties have realized the need to adapt their discourse and strategies to conform to the language and logics of modern bureaucratic resource management. This is a story of how whalers and antiwhaling activists have learned to think like the institutions they struggle against: to see whales like a state.

The Makah whaling conflict is best understood, then, as a story of translation, of the strategic adaptation of discourses and practices to fit a dominant interpretive framework. Telling a story of translation begins with what is being translated, and in the following pages I explore the identities, narratives, social institutions, and world views that are constitutive of Makah whaling and opposition to it. For Makahs, this conflict is about much more than staking a claim to interpretive sovereignty over the meaning of *the whale*. The need to defend treaty rights dominates local conversations on whaling. It is almost impossible to imagine a discussion of whaling in Neah Bay in which the 1855 Treaty of Neah Bay is not mentioned. Whales are but ancillary characters in such conversations, always playing instrumental roles, always preceded by their immense symbolic importance. Indeed, the whales in the Makah whaling conflict are ever in danger of being erased by their significance as much as by their insignificance. As the magnificent megafauna at the center of the whale-watching industry and as the formidable beings pursued in the historically and culturally meaningful activity of whaling, actual whales tend to show up infrequently in this story, and when they do, they seem to be already written over by many layers of interpretation. It *is* sometimes the case, as Pearson and Weismantel have noted, that the animals visible in the contemporary world "really *do* serve a function that is almost exclusively symbolic" (2010, 30). This conflict is as much about *the whale*—that abstract chimera that is neither singular nor plural—as it is about Yabis or any other singular living being.

One of the main arguments of this book is that we cannot understand the Makah whaling conflict without attending to its moral dimension—to the differing ideas about how humans ought to act with and toward whales and how those ideas align with the state's. Here, I join a growing number of anthropologists who recognize morality as a useful analytic category and legitimate subfield of inquiry.[14] My specific contention is that morality does not follow from ontology like a dependent variable, as is often suggested among anthropology's so-called ontological turn (see, e.g., Kohn

2013, Descola 2013).¹⁵ There can be moral consonance that spans ontological divides as well as moral dissonance that fractures ostensibly similar ontological viewpoints. Only by appreciating this can we understand the challenges and consequences of translating the disparate agendas associated with the Makah whaling conflict into something acceptable to federal fisheries managers. Only then can we have a full accounting of what is at issue and what is at stake in the Makah whaling conflict and grasp its likely future. By "what is at issue," I mean the actual disagreements that drive the conflict and make it seem intractable, as well as what has led to the conflict's current configuration and the strategies that the various actors have taken. By "what is at stake," I mean why the conflict matters to those involved; why people on both sides would dedicate years of their lives to a clash with no end in sight and with little promise of material or financial reward.

Where the mandala most accurately reflects the Makah whaling conflict is in its swirling, irreducible complexity: more than a political or moral or ontological struggle, the Makah whaling conflict is all of these things at once. When the whale approached the canoe, she approached a tangled mass of institutions and agendas, of ready-made narratives and interpretive frames. She approached a space where her intentions, her capabilities, her proper legal status, her proper moral status, even her genetic makeup had been discussed long before she, this singular living being, approached the canoe.

What whales *become* in the moments when they approach is not a matter of ontology pure and simple, not an instantiation of timeless, stable forms.¹⁶ Instead of *becomings*, which feel so permanent, we might think of these momentary points of contact as sudden eruptions of meaning-making set against the background hum of overlapping environmental imaginations. Yabis tended to matter only insofar as the qualities imputed to her threatened or advanced one narrative or another, or some version of *the whale* in its competing singular vision. Was she part of a protected subpopulation? Was she friendly? Was her body treated with respect? Were all of her tissues used or consumed? None of this lends itself to simple or reductive explanation, but that is rarely the case when nonhuman animals enter human spaces and become palimpsests of human dramas.

And this is a human drama, in the last analysis, albeit one with undeniable consequences for gray whales. As an ethnographic theory and method, decentering the human is among the worthiest of contemporary anthropological aspirations, but part of reaching for this goal involves recognizing that sometimes animals are marginalized figures in the stories that affect them the most. We find in reality that humans are often none too willing to give up their central positions. This is another unfulfilled aspiration of the mandala: that the whale, or any particular whale, would be at the center of a conflict with so many other parties vying for that place. What, in a different analysis, might have been understood as the agency of the whale—what whales are allowed to "say" and how they are allowed to say it—takes shape within systems of meaning-making and technocratic governance that are all too human. The most important question this book addresses, then, is not "What is a whale?" or even "What does the whale become?" but "What does the whale approach?"

The Makah Whaling Conflict

"Makah whaling conflict" is the term commonly used to describe the Makah Indian Nation's ongoing struggle to regain the domestic legal right to hunt gray whales and the varied efforts by antiwhaling activists to stop them. Like the whale that is ostensibly at its center, the Makah whaling conflict is a singular term with a definite article that stands in for a plurality of things: debates, protests, and legal battles, fought in courtrooms and conference rooms and on the high seas. It is a conflict that has taken place in fits and starts across twenty years and counting. Since 2000, a series of federal court cases and administrative proceedings have delayed the reauthorization of the hunt. A generation of would-be Makah whalers have come of age since then, while activists wonder how long their temporary victories can last.

The clash between Makah whalers and antiwhaling activists first came to public attention as early as 1994, when the Makah tribal government announced plans to revive its people's whaling traditions, and antiwhaling activists from across the region and the world announced their intent to stop them. The Makah are a small, federally recognized Native American tribe of about 2,600 members, around 1,200 of whom reside on the Makah Reservation in Washington State, at the northwesternmost tip of the contiguous United States.[17] The tribe's long history of whaling dates back to well before contact with Europeans. For as many as eight hundred years,[18] whaling reflected and reinforced the highly stratified social order of the Wakashan-speaking villages around Cape Flattery—whose descendants would comprise the Makah Tribe—providing chiefs or "big men" with the symbolic and material means to maintain and expand their social influence.[19] It is commonly said that in the 1840s Makah whale oil lit the nearby city of Victoria, then the largest settlement in the region. The 1855 Treaty of Neah Bay guaranteed the Makah Tribe the right of whaling and sealing in common with other citizens of the United States, the only Indian treaty expressly

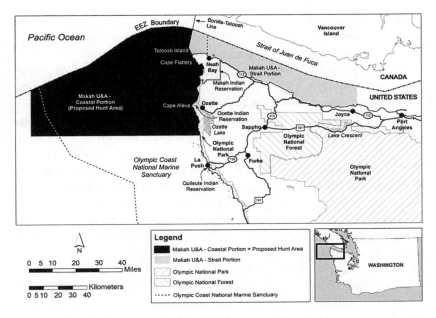

Figure 0.2 Map of area showing the Makah Reservation, Makah usual and accustomed fishing grounds (U&A), and the town of Joyce (home of activists Chuck and Margaret Owens)

to do so. The treaty included promises to outfit Makah whalers with materials and technology for whaling, reflecting the importance of the activity to the tribe and the region.

Despite the pride of place given to whaling in the Treaty of Neah Bay, by the end of the 1850s the importance of whaling for the tribe began to decline, and by the 1920s Makah whaling had ceased. A widely shared local narrative insists that Makahs voluntarily ceased whaling in the 1920s when they recognized that the gray whale population was in free fall. The strength of tribal affiliation in the Pacific Northwest reached its nadir around the same time due to assimilation pressures, trends in the availability of wage work, and intermarriage with white settlers and members of other tribes.[20]

The legal and political status of whales and other marine mammals changed dramatically in the half century following Makahs' cessation of whaling. In the 1960s and 1970s, whales became important symbols in the burgeoning environmentalist movement. The Marine Mammal Protection Act (MMPA) of 1972 banned commercial harvests of whales, dolphins, and all other marine mammals in US waters. Images of "charismatic megafauna" like whales, bears, and eagles were instrumental in the passage of the Endangered Species Act of 1973.[21] Even the International Whaling Commission, notoriously slow to realize that whale stocks were in decline, placed a moratorium on commercial whaling in 1982.

The same political climate that vaulted whales to international prominence and made them subjects of a new kind of moral concern also gave rise to new and resurgent forms of Native American identities. On the Northwest Coast, this resurgence was organized around the struggle for tribal fishing rights, particularly for the right to fish for spawning salmon in Washington rivers. The struggle culminated in the ruling in the 1974 *United States v. State of Washington*, known locally as the Boldt decision. The ruling entitled Washington treaty tribes to fifty percent of the commercial take of all salmon that passed through their "usual and accustomed fishing grounds."[22] Emboldened by the Boldt decision, in the early 1980s Makah commercial fishermen began exploring the possibility of a return to whaling. They viewed whaling and sealing as the logical next step in a vision of growing tribal sovereignty over resources.

In the late 1980s, Makah officials began furtively petitioning for the "delisting" of the Eastern North Pacific Gray Whale from the Endangered Species List. The stock was removed from the list in 1994. The tribe's announcement of their intent to resume hunting the species met with an immediate and predictable uproar. The Makah Reservation is just a four-hour drive from Seattle, in the proverbial backyard of some very powerful environmentalist and animal rights groups, where opposition to whaling is widespread and diffuse. Opinion polls conducted in the 1990s showed that over eighty percent of Americans opposed whaling under any circumstances, and there are reasons to believe that the proportion is greater in the Pacific Northwest, where images of whales are ubiquitous and where whale watching is a multimillion-dollar industry.[23]

At the behest of the National Marine Fisheries Service (NMFS), Makah tribal leaders agreed to first seek approval for the hunt from the International Whaling Commission (IWC). Since the 1970s, the IWC has granted special exemptions for aboriginal subsistence whale hunts on whale populations that would not support a commercial hunt. The US delegation placed a proposal for the Makah hunt before the IWC at the 1996 annual meeting in Aberdeen, Scotland, but the proposal met with criticism. While the Japanese delegation enthusiastically supported the request, several member nations questioned whether the tribe had a continuing subsistence need given its long hiatus from whaling.[24] The proceedings were thrown into an uproar when a group of Makah elders, who had flown to the meeting with anti-whaling activists, spoke out against the hunt in front of the full assembly.

Delegates from antiwhaling nations questioned how the hunt could be considered cultural if the tribe's own elders opposed it. It was clear that the IWC's institutional concept of culture equated legitimacy with unanimity.[25] The US delegation withdrew the Makah proposal. The following year, however, the United States secured an annual harvest quota of five gray whales for the tribe (in what was widely described as a backroom swap with the Russian delegation; see Ziontz 2009).

The events at the IWC meetings in Aberdeen left an indelible mark on the collective memory of the Makah community and especially on the tribal officials who had experienced them directly. It would be simplifying things only slightly to say that Makah leadership responded by doubling down on a narrative of cultural uniformity. The need to present a unified front also led to overt attempts to silence dissenters in the community, all in the interest of seizing control over an ostensibly unified and authorized Makah point of view.

As the scheduled date of the approved hunt grew nearer, dozens of journalists occupied Neah Bay for weeks on end. The tribe held daily press briefings to satisfy public interest.[26] Makah officials and activist leaders were frequent guests on national talk shows. Freelance journalist Robert Sullivan spent considerable time on the reservation over the two-year period from 1998 to 1999, documenting the hunt and its buildup through his acquaintance with whaling captain Wayne Johnson. Sullivan later wrote a book-length chronicle of the hunt, told mostly from the perspective of Johnson and his close associates.

Beginning in the fall of 1998, local antiwhaling activists held weekly protests near the reservation boundary on the only road leading into town. The Sea Shepherd Conservation Society, an antiwhaling organization formed by disgruntled Greenpeace cofounder Paul Watson (now of Animal Planet's *Whale Wars* fame) anchored *Sirenian*, a repurposed Coast Guard cutter painted black in the Sea Shepherd style, outside of the Makah harbor for several months beginning in late summer. When a tribal whaling crew finally took to the water, they were accompanied by a flotilla of antiwhaling protesters. They were also joined by a consortium of whale-watching companies from nearby Victoria, British Columbia, who used their private fleet in an attempt to disrupt the Makah hunt. In response, the US Coast Guard established a "moving exclusionary zone" around the whaling canoe. Any protest craft that entered the zone was quickly confiscated. By the third day of the hunt in the spring of 1999, most of the activists' vessels had been seized.

The 1999 hunt was filmed by a news helicopter and broadcast live on regional news networks. Many Makahs watching at home on their televisions reported being moved to tears when the whale was harpooned. The whale was towed by a Makah commercial fishing vessel around Cape Flattery to Neah Bay harbor, where it arrived on the beach almost twelve hours after the kill. The ship detached itself and allowed the canoe to tow the whale the final distance across the harbor to shore, aided by canoes from other tribes in the region, who participated in a show of tribal solidarity.

The hunt resumed in the spring of 2000, and antiwhaling activists returned with new tactics. In order to avoid having their larger vessels confiscated by the Coast Guard, activists used personal watercraft launched from the deck of larger vessels to try to interfere with the canoe. No whales were killed before the Ninth Circuit Court temporarily suspended Makah whaling in June of that year pending further regulatory review. Activists had filed a lawsuit alleging that NMFS had violated the National Environmental Policy Act by preparing an environmental assessment of the tribe's proposed hunt *after* approving it rather than before. The court ordered regulators to take a "hard look" at the potential environmental impacts before returning with a revised assessment.[27] A second lawsuit led to a separate ruling in 2002 that called for a full environmental impact statement (EIS) and also ordered the tribe to seek a waiver of the Marine Mammal Protection Act.[28]

Citing frustration over these regulatory delays, five Makah whalers from the 1999 crew hunted and killed a second gray whale in 2007 without the approval of the US federal government. The incident came to be known locally as the "rogue hunt." The men harpooned and shot the whale from two powerboats but did not immediately kill it. One of the whaling guns was lost when it flipped overboard, and repeated shots to the head from a smaller rifle failed to kill the whale.[29] The US Coast Guard intervened and apprehended the crew before a replacement gun could be brought to the scene. The whale died of its injuries nearly ten hours later as it drifted out to sea.

There were whispers that the rogue hunt had been unofficially encouraged by members of the tribal council as a test of whether the Treaty of Neah Bay could stand up to a legal challenge in the absence of federal regulatory approval. If that was indeed the purpose behind the hunt, the answer was resounding. The Coast Guard stopped the hunt and confiscated the whalers' boats and equipment, and two of the whalers served sentences (of five months and 90 days, respectively) in federal prison for the offense, including Wayne Johnson, the captain of the 1999 crew. The state's overwhelming authority had already been established for the activists a decade earlier, when protester Erin Abbott's personal watercraft was run over by a Coast Guard Zodiac as she entered the moving exclusionary zone around the Makah canoe during the 2000 hunts.

I first arrived in Neah Bay in 2010, just in time for a major development in the debate over the reauthorization of the Makah hunt. The results of a study on the population genetics of the gray whale had shown unexpected differentiation in mitochondrial DNA haplotypes of certain stocks, causing NMFS to halt work on an EIS that had been in progress for years. NMFS assembled a task force on gray whale stock identification that convened in 2013. At the time of this writing, all sides await the publication of the final EIS, which will include the agency's recommendation on the future of the Makah hunt.

The Other Leviathan

Unlike the whale that was killed in the 1999 hunt, the whale that approached the two Makah powerboats during the unauthorized hunt in 2007 had been given a name before that day. The nonprofit Cascadia Research Collective, based in Olympia, Washington, had dubbed the whale CR-175. "It turned out to be a fairly well-known whale that was killed," Cascadia's cofounder John Calambokidis told a local reporter after the rogue hunt. CR-175 had been spotted 143 times along the Pacific coast between Northern California and Vancouver Island in the previous twelve years, often feeding alongside Freedom, a whale that a local activist group had named and adopted through a program run by Calambokidis.

Cascadia's cataloguing efforts form part of a broader regime of tracking, quantifying, and analyzing gray whales that informs federal cetacean conservation policy. The organization is particularly focused on a subset of gray whales that frequent the Puget Sound. The Eastern North Pacific gray whale has one of the longest migrations of any animal on Earth, moving from winter calving grounds in Baja California to summer feeding grounds in the Bering and Chukchi Seas, but scientists have known for some time that not all of the estimated 20,000 gray whales in the larger population travel all the way north every year. Some two hundred gray whales forage during the summer months between northern Oregon and southern Alaska, an area that includes the Makah Tribe's usual and accustomed fishing grounds. This subpopulation is now known to federal scientists as the Pacific Coast Feeding Group (PCFG). They were known at the time of the first Makah hunt as "summer resident whales."

Federal scientists have been unable to agree on whether the PCFG warrants consideration as a separate management unit. The 1994 decision to remove the Eastern North Pacific gray whale from the Endangered Species List made no mention of them, nor did the Makah Tribe's letter of intent or the quota they received from the International Whaling Commission. The first environmental assessment of the Makah hunt published by the National Marine Fisheries Service in 1997 mentioned the residents only briefly, and the Makah Tribe's marine biologist on staff denied their existence until as late as 2007. It was not until antiwhaling activists made the summer residents the centerpiece of their successful 2002 lawsuit against the US Department of Commerce that the PCFG became the object of sustained scientific scrutiny. In 2012, NMFS canceled the EIS it had been working on for years because of new scientific data on this putative stock.

It would be difficult to overstate the role that regulatory science has played in the Makah whaling conflict. The state is ubiquitous in the region and in the area around the Makah Reservation in a way that perhaps only someone who has spent time in the Pacific Northwest may fully appreciate. No fewer than six different federal agencies have a presence on the reservation, and if you look at the narrative of the hunt above you can see evidence of the state's ubiquity on this issue. The whaling crew in the 1999 hunt had to wait until the federal observer was present before they could harpoon the whale. The activists were forced to look on helplessly during the successful hunt because the Coast Guard had confiscated most of their vessels for violating the Marine Mammal Protection Act. The magic bullet that brought an indefinite halt to Makah whaling in 2002 had to do with the minutiae of federal environmental policy regarding the PCFG subpopulation. The 2007 rogue hunt ended when it did because the Coast Guard apprehended the crew for hunting without the permission of the federal government. None of this is to mention the role of the state in bringing the Makah Tribe into being as a juridico-political entity in 1855, promising it the right to whale in perpetuity, serving as the significant other in the tribe's long history of struggle against the state of Washington over fishing rights, and playing the role of arbiter in the tribe's ongoing quest to recapture resource sovereignty.

When the state appears in this narrative, as it does with increasing frequency in this book's second half, it does so primarily through the indexical marks it leaves upon the story. This is not a story of the state or its agents but of a conflict shaped and constrained by the state's moral economy, and of individuals placing the state's preferred signs and its values into novel combinations for strategic ends. The state, as anthropologists have shown, is not one thing but many and is not static or fixed (Gupta 2012). Yet for the people most affected by the state—those whose aims and actions are encompassed by it—the state often really *does* have the appearance and character of a monolithic agent, even if it is, in the last analysis, no more than a heterogeneous accretion of mostly unspectacular actions. Patchen Markell

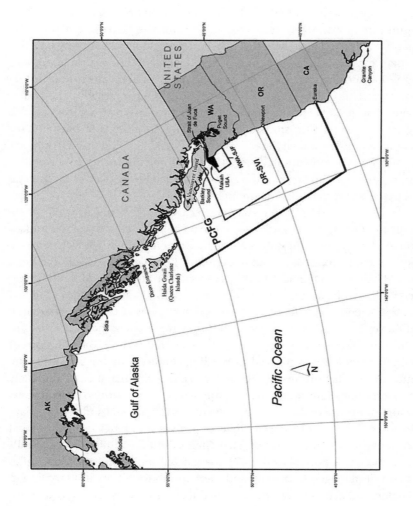

Figure 0.3 Estimated range of the Pacific Coast Feeding Group (PCFG), or summer "resident" whales

(2003) calls this the state-effect, and it is itself a crucial part of how the state is produced and maintained.

The state's most common manifestation in this case is the National Marine Fisheries Service, commonly referred to by its acronym, NMFS, the pronunciation of which ("nymphs") adds to the general perception of it as something abstract and ethereal. A division of the National Oceanic and Atmospheric Administration (NOAA) within the Department of Commerce, NMFS is charged with such diverse responsibilities as setting fisheries quotas, measuring seawater chemistry, modeling marine mammal populations, monitoring shipping traffic and other potential threats to marine life, and representing the United States at meetings of the International Whaling Commission. Even whalers and activists who are on a first-name basis with agency personnel often speak of NMFS as if it were a singular thing with inscrutable motives and endless means. From the functional opacity of much of the agency's decision-making to the deadening impotence of stakeholders' frustration with some of the agency's decisions, the decrees of the state seem to issue forth as if from on high.

When a US Coast Guard vessel ran over a protester during a Makah whale hunt in 2000 after she entered the exclusionary zone around the Makah canoe, the Coast Guard sailors allegedly warned her two fellow activists—both trained paramedics—not to approach her to render aid, or "the same would be done to them." Clearly, the spectacle of the scaffold has not fully receded from how the state presents itself to dissidents. These are the features that tend to be left out of accounts that emphasize the dispersed nature of power in disciplinary regimes, and they represent some of the ways in which the state is not like any other character in this story. The eruptions of legitimate violence in the Makah whaling conflict belie the sense that the bureaucratic institutions of the modern state legitimize themselves solely through nonviolent means such as the perceived authority of scientific expertise or the use of discursive barriers to entry.[30] Whalers and activists have accepted the need to translate their objectives into what I call the moral economy of NMFS because of what Glen Coulthard euphemistically calls "non-discursive power" (2014, 47): the state's political and economic influence and managerial capacity and, perhaps above all, its military might. Whalers and activists have each learned that if they defy federal regulators, the state may show up to their next encounter in a more threatening form. If the Coast Guard were not a phone call away, the story of the Makah whaling conflict would look very different.

As the battle over Makah whaling has shifted to courtroom dissections of federal fisheries policy and to public-forum discussions of conservation documents, a shared sense has settled in among the parties to the conflict that the state will decide which facts about cetaceans matter and which do not. The technocratic barriers reemerge.

NMFS's approach to managing whales, like its approach to managing fish and other commercially important "living marine resources," is based on periodic assessments of each whale "stock," which are used in turn to determine the potential biological removal, or PBR, of a given stock at any given time. This process is the same for bottlenose dolphins as it is for minke whales and differs from the management of fish species in only minor ways.[31] The formula looks something like this:

$$\text{minimum estimated population} \times \text{maximum annual growth rate} \times \text{recovery factor}$$

For instance, NMFS calculated the PBR of the critically endangered Western North Pacific gray whale, with a population estimated to be around 135, at .06, or one human-induced fatality (or "take") every seventeen years.[32]

The question of how human beings ought to act in relation to a given whale stock is thus reduced to a single number: the PBR. Lying outside of this abstraction are virtually all of the moral concerns that brought antiwhaling activists to the Makah whaling conflict in the first place, as well as many aspects of Makah ontologies of whales and whaling, whether spiritual or economic or cultural. Lying outside this abstraction are most of the enactments of Yabis I described above. As an obligatory passage point for citizen participation in the Makah whaling conflict, the conservationist ontology of US federal fisheries management thus constitutes an ethical as well as an ontological stricture, assuring that translation will have consequences.

In the pages that follow, I may seem deeply critical of the state and its system of stock-based fisheries management, but it is not my intention to criticize the important work done by conservation scientists around the world. A Makah commercial fisherman once told me he filled his entire yearly quota for one target fish in a single day by scooping up an entire school of fish; it is not difficult to imagine what the fishery would look like with such technological capabilities and no catch restrictions. Whether the remarkable recovery of many large cetacean species since the middle of the twentieth century can be attributed to conservation measures is less clear. The ostensibly conservation-oriented scientists and administrators at the International Whaling Commission equivocated over the decline of the Southern Ocean whale fishery until the 1960s, long after what historian Graham Burnett (2012) calls "world opinion" had shifted in favor of preservationism. The moratorium on commercial whaling that is currently in place at the IWC runs against conservationist principles in a number of key respects, since multiple whale stocks have since shown signs of recovery and some, arguably, are overpopulated, which has been the justification that whaling nations like Japan and Norway have used to push for a "revised management plan" that would allow commercial whaling. It is entirely possible that commercial whaling could be in line with sound conservation principles. It is significant, then, that the ethos of conservation—the preservation of species and stocks in "sustainable" numbers—precludes, in practice, many of the moral arguments that saved the whales in the first place.

Ethnography in a Dispute Setting

This book explores a controversial topic, one where even the most basic generalizations can be fraught with political implications. To speak of *the Makah*, for instance, while unavoidable at times, can be deeply misleading if done carelessly or deliberately to conflate the different senses of the term. As a proper noun, *Makah* can mean a number of distinct things that social scientists should take care not to confuse. The Makah Tribe is a legal and political entity, a quasi state with a measure of sovereignty as a domestic dependent nation, and also something of a going concern.[33] *Makah* also commonly refers to a place, the Makah Reservation, and a people: a collection (but not always a collective) of individuals and factions who, like the members of any community, align themselves with and against one another in various ways, along familial and political and ideological lines. The tribe consists, moreover, of a postcolonial conglomeration of five semiautonomous villages, a fact that each day tests tribal solidarity and leads to real disagreements over matters that the tribal government has deemed to be of public importance.[34] Makahs have various levels of investment in the very notion of the Makah Tribe as such.

Even among Makah supporters of whaling, there is no single "Makah" opinion. The pursuit of whaling rights means markedly different things to different people. For some, whaling is an effective form of protest against the state; for others, it is a way to leverage good working relationships with the federal government to secure more resource rights for the tribe. For some, whaling is materially important as a source of healthy protein; for others, it is of mostly symbolic importance (it is generally admitted that gray whale meat does not taste very good). For some, whaling is a tribal venture; for others, it is best left to extended families to organize among themselves. Makah whaling seems at once to be a way to define and distinguish Makahs from whites as well as a means of distinction from other Indians in the region, charting a kind of Pan-Indian identity and solidarity alongside a version of tribal exceptionalism. Religious practices associated with Makah whaling are also characterized by their variability, which I explore at length in chapter 4.

In a similar way, antiwhaling activists in this conflict are united by little beyond their shared opposition to whaling. Of the groups involved at the early height of the conflict in the 1990s, some were international organizations with global ambitions, and others were single-issue local protesters with a "not in our backyard!" mentality. Some were vegans who opposed the killing of all animals; others ate beef and pork and saw no contradiction in opposing whaling. No one I worked with closely on the activist side was obviously motivated by racialized resentment; indeed, the local activists with whom I spent most of my time worked actively to counter and distance themselves from what they perceived to be racist arguments. Racialized hostility and racist undercurrents were certainly present in the broader reaction to Makah whaling across the region, however, as well as in the many threatening calls and messages the tribe received around the time of the 1999 hunt (Erikson 2002).[35]

Little of this complexity has made its way into previous accounts of the Makah whaling conflict. A key fault line lies between those accounts that question the existence of a unified Makah point of view and those that affirm the rhetorical fiction of a monolithic Makah Tribe acting with a single will and purpose.[36] Makah whalers and tribal leaders tend to come across rather flatly in some of these accounts, where one group of tribal elites speaks for what "counts" as Makah, and whalers are portrayed as neotraditionalists seeking to revive or "reaffirm" cultural identity (e.g., Miller 2000) rather than as savvy political actors in their own right.

The activists have generally come off much worse. I am frankly shocked by the number of accounts that have explained all antiwhaling activism in the Makah whaling conflict as an expression of racism, apparently without bothering to speak to a single whaling opponent (e.g., Coté 2010, 150; Marker 2006; Barton 2000). The historian Joshua Reid (2015, 3) could not be plainer on this point. "These critics were racist," he writes, gathering all those who opposed the Makah hunt under a single heading, "and they were also wrong." Conflating local protesters and the Sea Shepherd Conservation Society with all of those who wrote angry, racially tinged letters to the editor of the *Seattle Times*, Reid writes, "[These critics] have overlooked—and continue to ignore—the historical and cultural connections Makahs have to the ocean" (3).

I can understand the need for strategic essentialism from the point of view of the Makah tribal government, especially given their experience with the International Whaling Commission. When the Alaska Eskimo Whaling Commission, veterans of the IWC process, heard about the episode at the IWC, they publicly criticized the Makah for failing to present a united front. "The Makah did not do their homework," an Inupiat spokesperson told the *Seattle Times*. "There's too much controversy to have objectors in the tribe. The Makah must begin to speak with one voice, or they are sunk."[37] Moreover, of the two main sides to this conflict, only one has to deal with questions of *authenticity*. Even if cultural exemptions have benefited the corporate interests of Native Americans in many instances, they subject tribal governments to the caprice of the exemption-granting institution and the particular way it understands *culture*, which, as Raibmon (2005) tells us, tends to be along the lines of stereotypes of early twentieth-century anthropological conceptions of culture as fixed, uniform, and homogeneous.[38] Political exigencies thus often seem to require that Makah leaders treat the multiple meanings of *Makah* as if they were one, which entails the denial of differences of opinion within the community. It is an unfortunate situation, and one for which anthropologists are far from blameless. Surely, we can recognize our own concept of culture (albeit twisted nearly beyond all familiarity) at the center of this politics of recognition. But I, for one, cannot in good conscience reproduce narratives that did not happen in the service of the political interests of some people within the tribe, much less if it comes at the expense of my other informants.

The political scientist Claire Jean Kim (2015) has observed how conflicts such as this, where concerns about animal ethics and social justice intersect (she had the Makah whaling conflict specifically in mind), can lend themselves to partisan misreadings, allowing scholars from various fields to present oversimplified morality tales to audiences eager to take one side or the other. Kim offers the example of Rob Van Ginkel's essay "The Makah Whale Hunt and Leviathan's Death" (2004), in which, despite announcing his intention to "inventory" the Makah whaling debate, Van Ginkel presents a sympathetic portrait of Makah whaling while barely hiding his contempt for the antiwhaling activists, whom he accuses of anthropomorphizing whales. In her analysis of the Makah whaling conflict based on a reading of publicly available documents and secondary sources, Kim resists the temptation to oversimplify the conflict, instead calling for what she refers to as a "multi-optic approach" that encourages an ethic of "mutual avowal." Kim offers a set of recommendations by which stakeholders and observers might take the claims of all parties seriously and respectfully without necessarily changing their opinions.

This book shares the goal of minimizing avoidable bias and rendering all parties sympathetic and understandable. In place of Kim's normative framework with its a priori commitment to detecting and weighing competing oppressions, I substitute a descriptive approach that treats the conflict as more than just a contest of moral assertions. I take up Didier Fassin's (2008) call for an anthropology that treats morals as descriptive objects, as opposed to a moral anthropology in which anthropologists become moralists themselves (334–35).[39]

The descriptive approach entails some commitment to ethnographic neutrality, even if it is something to be endlessly strived for and never fully attained. I find that the most practical means of pursuing ethnographic neutrality is to refuse to settle moral or political questions in advance—to refuse to say that all activists are motivated by racism, for instance, or to refuse to declare a priori that Makah whaling is entirely an act of cultural revitalization. Careful use of terminology is also important, like refusing to refer to whales unproblematically as "resources," or taking care not to refer haphazardly to the collective will of "the Makah." Mostly, it means resisting the urge to see one party or another as on the right side of history.

The challenge of doing an ethnography of a conflict is to do justice to both sides by avoiding this very favoritism. To avoid making caricatures of one side or the other requires getting to know them, which in turn requires their generosity and hospitality. I have been the beneficiary of such hospitality many times over, but my determination to be fair and honest and neutral with the concerned parties precedes and exceeds the reciprocal obligation created by that hospitality. I found that most of my informants on both sides of the issue accepted and even invested in my role as a neutral scholar.[40] To give an intentionally biased or slanted portrayal would, among other things, betray the trust they placed in me.

It may help to clarify that I have no interest in speaking for the Makah, or any Makahs, or antiwhaling activists, or whales. All but the last are perfectly capable of speaking for themselves. My aim instead is to speak *of* the Makah whaling conflict and the parties to it; the narratives, values, and social institutions that are implicated in it; the shape it has taken and what it can tell us about other conflicts of this kind. All of this can be understood from a place John Searle (2006) calls "epistemic objectivity" or "observer-independence," or at least something approaching it.[41]

It is here that gray whales have their greatest say in the matter. Because they are beings that traverse national boundaries on their annual thousand-mile migration, it cannot reasonably be said of gray whales that they belong to any particular human group.[42] They are no one's patrimony. As such, the movements of gray whales provide a warrant to cross the ethical boundaries that some scholars would see us erect between peoples, or what Michael Brown calls the "benign form of quarantine" (2009, 209) that has emerged in legal discussions of cultural heritage. Gray whales' migratory habits demand that we cross these borders, giving us a shuttle into the quarantine zone.

The Ethnographer

When I say that I do not want to produce an intentionally biased or slanted portrayal, that does not mean that I think I have produced a text that is fully free of either. You would find few anthropologists today who believe such a thing to be possible. Nor do I lack political leanings myself. While I am encouraged by Edward Said's statement in the first few pages of *Orientalism* that a text can be less, rather than more, partial than the individual that produces it (1978, 10), I do not think that perfectly applies to this situation, either. My political leanings have never been so fixed or coherent as to provide a stable reference point from which to make my analysis less partial. As a child, I remember balling up my fists in anger over the whale-killing scenes in a movie adaptation of *Moby-Dick*. Although I'd never seen a whale in person, I remember thinking of them as giant, gentle, alien-like intelligences that we ought to try to communicate with rather than kill. But I also grew up hunting deer in rural Michigan with a compound bow and later with a firearm; and although my enthusiasm for hunting has tailed off in recent years (even after being slightly rekindled by my time in Neah Bay), I remain an avid fisherman. My feelings about whales had already become more nuanced by my mid-twenties. Once in the field, I found that my personal political convictions varied depending on where I was and how long I had been there. When I had been on the Makah Reservation for six months, the thought of killing five gray whales in a year seemed completely reasonable. The image of a dead whale "on the beach" even started to look like the whale's most natural form to me. When I was interviewing activists, and especially when I was their houseguest for several days or more, I found myself feeling less favorable toward whaling, and I even felt the old activist rage well up in me from time to time, although never in the specific direction of Makah whalers, whom by then I knew too well to judge very harshly. A reader might see this vacillation as moral weakness, but I would rather see it as a testament to the power of context and human relationships. Had I been able to get to know any gray whales on a personal level, this might have been a different story.

While I divided my time between two main groups for this research and do refer to the two "sides" of the conflict, Makahs and antiwhaling activists are too differently constituted as groups for this to be a true comparison. The members of the Makah Tribe who live on or near the Makah Reservation are linked by a sense of place as well as shared histories and kinship ties, but they are not unified in their opinions on Makah whaling or any other issue (except, perhaps, the importance of treaty rights). Antiwhaling activists, in contrast, are linked by their shared opinion on a single issue and sometimes little else. No one is born into a community of antiwhaling activists. The realities of working with each group differed accordingly. The activists I focus on in this book are mostly local residents of the Olympic Peninsula, but they were joined by protesters from across the country who traveled to the region for the single purpose of opposing the Makah hunt. Interviewing activists meant visiting their homes and sometimes staying over as a houseguest, whereas in Neah Bay I had my own apartment most of the time.

Although I cast a wide net, so to speak, and I lived on the reservation for a year in total, the methodological heart of this project is the ethnographic interview, each lasting one to three hours, of which I conducted dozens. I used a flexible interview "script" that I adapted for each interviewee, a set of topics that I knew I needed to cover at some point in the interview. The conversations were otherwise person-centered and unstructured.[43] Participants were selected through networking and word of mouth. Almost everyone I spoke with on the Makah side had some connection to "the whaling," as it is often referred to locally. Because of the nature of extended family affiliation and its role in the transmission of cultural and religious knowledge in the community (more on this in chapter 2), I deliberately chose interviewees from a broad cross-section of families. I also noticed that when doing an ethnography in Neah Bay, one is continually directed to the same roster of experts by topic: Darrell Markishtum and Greig Arnold on "the spiritual side of things," for instance, or Micah McCarty and Greg Colfax on just about anything. I made sure to include a number of interviewees who were not on these informal lists. Because of the media attention this conflict has received, some of the Makahs who feature in this narrative might already be known to readers.[44]

I conducted fewer in-depth ethnographic interviews per capita with more Makah tribal members: around forty interviews with about twenty-five Makah interviewees, versus fifty hours of interviews with fewer than ten antiwhaling activists. Much of my analysis focuses on just two important figures on the activist side. As a result, I got to know Chuck and Margaret Owens very well, and my narrative and descriptions of them might show it.

I got to know a number of Makah tribal members at least as well, including my hosts for several shorter visits, but I have deliberately kept the latter out of this narrative. I lived in Neah Bay for nearly a year in total, and I needed backstage friendships where I could enjoy not being an ethnographer for just a few minutes of the day.

An anthropologist who conducted fieldwork in Neah Bay around 2000, Patricia Erikson, noticed that she was continually bumping into the legacies of previous generations of anthropologists as she went about her fieldwork. Memories of them seemed to affect the way local people interacted with her.[45] I did not feel this way, although I did notice that there was what seemed like a prepared role for anthropologists in the community, what we might call the anthropologist slot.[46] For some local people, my presence elicited what seemed like a practiced combination of patience and caution. These individuals manifested a deeper practical understanding of why I was there than what one might expect to find in other field settings. Mostly, though, my interactions with people were pleasant, gratifying, and at times exhilarating, from three-hour chats with Micah McCarty at his home on the bank of the Wa'atch River to long, contemplative moments spent listening to Greg Colfax in his front yard as he reflected on what he knew about the life of "long ago." I can remember sifting through binders of newspaper clippings with Margaret Owens for hours on end, feeling energized by her boundless passion for the issue and thinking that if I could just somehow miraculously *know* everything she remembered about the conflict, I wouldn't have to do a bit more fieldwork.

The bulk of the research and interviews on the Makah and activist sides were conducted concurrently. For eight months in 2011, I lived in a studio apartment on the reservation owned by a Makah woman and her husband, a white commercial fisherman. I frequently had morning coffee with a group of men that included several whalers, and I spent many days digging clams or working on other small projects with Wayne Johnson. I regularly drove the ninety minutes to the Port Angeles area to meet and talk with activists. In all, I spent far more time with Makah informants than with activists, and I got to know far more Makahs (and prowhaling Makahs) than any other group.

The Setting

The Makah Reservation is a forty-six square-mile section of land at the northwest tip of Washington's Olympic Peninsula. The terrain of the reservation is rugged and beautiful, encompassing sandy beaches and broad river mouths and the scenically stunning Cape Flattery. From the beaches the land slopes quickly up to the forested peaks of the peninsula's rocky interior. The reservation is roughly bisected by Wa'atch Prairie, a narrow, low-lying plain that was in very recent geological time the seafloor of a channel that connected the Pacific Ocean to the Strait of Juan de Fuca.

Neah Bay is the primary population center of the Makah Reservation, the site of most community and "cultural" events, and the reservation's only town.[47] The village is arranged in a crescent around its formidable harbor, which is filled, in the cool summer months, with enthusiastic non-Native sport fishermen and, in the stormy winter months, with the sleeping giants of the tribe's large and successful commercial fishing fleet. Several rows of houses—most of them single-story modular and mobile homes—reach back from the beach, home to about 850 of the reservation's residents. Washburn's General Store, the Makah Mini-Mart (the only profitable tribal business in town, or so the rumor goes), and the community hall are key landmarks, along with at least five Christian churches. On the main street is a popular espresso stand open only in summer months. Potlatches, funerals, and basketball tournaments are held at the new tribal gymnasium on the main street, one of the centers of civic life. The gym's location in the business district sparked controversy when it was built, as did concerns about the source of its funding.

Not far down the street from the gym, an aging wharf juts into the harbor where most tribal commercial fishermen offload their catch. At least a few times each day, and almost constantly during peak seasons, a small fleet of forklifts shuttle huge crates of fresh-caught fish into the gaping maws of idling trucks that wait near the wharf's entrance.

At the east edge of town is the US Coast Guard Station, sitting roughly across from the Makah Cultural and Research Center (MCRC). Two giant, wood-carved statues of a Makah man and woman in traditional regalia stand in front of the MCRC to greet tourists as they enter the reservation, arms outstretched and palms facing upward in the traditional pose of the Makah welcome dance.

Until the 1930s, the Makah Reservation was accessible only by sea. Today, the only overland route may be just as treacherous: nineteen miles of serpentine highway that run along the Strait of Juan de Fuca, gouged into cliffs of sheer rock. All along the route, lichen-covered trees loom above the highway as if threatening to overtake it—and at least once every winter, they do, as the cliffs above give way and mudslides a quarter mile long envelop

Figure 0.4 Statues at the Makah Cultural and Research Center
Photo by Gloria Wadzinski

the road. Visiting journalists are given to describe the road with grandiose adjectives, and understandably so. The road serves as a fairly obvious metaphor for the way that Makahs living on the reservation view their connection to the rest of the United States: a thin, tenuous thread that could at any moment break, but is at present a conduit for the lifeblood of the community. To know this road, in fact, is to know much about the local conception of terrestrial space. The people of Neah Bay traverse the road constantly, some every day, zipping around each hairpin turn with what looks to the outsider like reckless abandon. There is a sense that no one is a resident of Neah Bay or the Makah Reservation exclusively; rather, they are more adequately described as citizens of the region, thinking nothing of leaving in the morning for Port Angeles (the closest nearby city, two hours away) or Silverdale or even Seattle (four hours, a ferry, and nearly 300 miles away) in the morning with a nonchalant "see you tonight." Many have a family member living in Port Angeles with whom they can stay if need be. A few have second homes there themselves.

Down that road from Neah Bay, about ninety minutes from the reservation border, Chuck and Margaret Owens reside in the woods near the tiny town of Joyce. During the height of the Makah whaling conflict in 1999, Chuck and Margaret and their ad hoc antiwhaling group (which consisted mostly of elderly women) would traverse the road each weekend to set up protest roadblocks near the Makah Reservation border. During the last few months of my fieldwork I traveled the road with about equal frequency to visit them, as well as to obtain supplies from Port Angeles. It never occurred to me until afterward that the road was anchored on both ends by museums: the Makah Cultural and Research Center on one end, and the Joyce Historical Depot and Museum—where Margaret Owens serves as a tour guide and curator—on the other. Occasionally, I would meet Margaret at the Depot to talk about whales and whaling.

The above description of the places involved in this research, like all such descriptions, is liable to be an evanescent one. Shortly after my fieldwork, a new fishermen's wharf was built that would support the weight of tractor trailers, obviating the forklift's busy play, such a feature as it was of my memory of the place. This is a useful reminder that the analysis in this book is based on a particular moment in time—between 2010 and 2012, mostly—when the future of the Makah hunt was uncertain and the memory of the conflict's first crescendo was still fresh in everyone's mind.

The spatial metaphor I have invited the reader to consider, *what the whale approaches*, may suggest a kind of stability and timelessness that is not at all the case. This book is, above all, a story of adaptation and concession, of the strategic mobilization of established categories and the multidirectional impact of those strategies on actor and category alike. For the tribe, the meaning and purpose of contemporary whaling shifted before it even entered the public eye. Born of commercial aspirations and a vision of growing tribal resource sovereignty, Makah whaling became a politicized and self-consciously cultural project that utilized the "ceremonial and subsistence" category originally won through lawsuits over commercial fisheries. This part of the story involves a substantial amount of contingency, almost none of which makes it into official accounts of the hunt.[48] Meanwhile, even as the whale swam toward the Makah canoe in 1999, a lawsuit was in federal court that showed the activists were slowly adapting to the framework of federal fisheries management. Their subsequent efforts along these lines have doubled down on this point, as they accept the moral and ontological suppositions of a technical framework that is radically different from their own, tacitly affirming a way of categorizing whales that runs against everything they have said and done with regard to their motivations, but which they feel might be the only way of stopping the hunt.

Chapter Outline

This book can be roughly divided into three parts. Chapters 1–4 focus on what might be called the Makah side of things, with chapters 5 and 6 focusing on the activists, although in practice and at key points in the text the two narratives intertwine. The final chapter in each of these two parts (chapters 4 and 6) considers the way the moral economy of NMFS works with and/or against the world views and enactments of gray whales described in the preceding chapters, reflecting on the attempts that actors on each side have made to translate their aims into the state's authorized discourse on whales and whaling. Finally, chapter 8, the sole chapter of the book's third part, returns to the topic of the whale's approach in order to explore its interpretation as part of the broader enactment of gray whales by the different parties to the conflict, beginning with parallel narratives offered by two Makah whalers who participated in the 2007 rogue hunt.

Any adequate account of the Makah whaling conflict must begin by complicating the received narratives that have taken hold in the press and popular imagination: to wit, that the hunt represented a *reaffirmation* of identity and sovereignty and an exercise in cultural self-determination. Such explanations fail to account for the fact that a substantial portion of the tribe did not support whaling when the hunt was first announced, only to coalesce around the initiative once outside resistance from antiwhaling activists was met. Such accounts also fail to explain the tenacity and commitment shown by the tribal government and its constituents in reviving whaling despite the fact that it arguably only benefited a fraction of the community (those with ancestors who were whalers). Chapter 1 situates the revival of the hunt in historical context, showing how contemporary Makah whaling is best viewed as part of a decades-long political struggle that is complex and, in several ways, contingent.[49] Providing a brief history of Makah whaling from its role in precontact society through "the fish wars" of the 1960s, I show how the struggle for resource rights became entwined with the (re-)emergence of Makah identities in the twentieth century. When new battles over marine resources arise, those identities become part of the stakes alongside older forms of social distinction predicated on inherited social status. Thus, originally born of commercial aspirations and a growing sense of resource sovereignty, whaling has become part of a narrative of tribal unity and simultaneously a means for some actors to (re-)assert traditional hierarchies and privileges.

Whaling is a political project that depends upon the moral presupposition that whaling is acceptable, something that is often expressed in local terms with the claim that whales are edible, or things-to-be-eaten. The utilitarian animal ethics contained in the category of edibility are so widely shared in the community that few Makahs question it, even among the opponents of whaling within the tribe. Reinforced by the circulation of images and motifs like "a whale on the beach," the edibility of whales flattens idiosyncratic differences between the various enactments of gray whales within the community. Some Makahs view whales as unique and special, whereas others view whales as hardly different from salmon except in size, but almost everyone agrees that whales are things-to-be-eaten. This moral sentiment diverges from the majority view of whales in the United States. Keenly aware of this, Makah supporters of whaling rarely miss an opportunity to emphasize their beliefs about the edibility of whales, whether on social media or in private conversation with a visiting anthropologist. Often, this purposeful provocation is expressed using some variation of the same single-word exclamation in response to a question or comment or image of gray whales—"Yum!"—and is intended to shock or offend a perceived animal-loving public. Chapter 2 explores what I call the ethics and politics of *yum*. I examine what it means to emphatically announce the edibility of whales in an interview with an ethnographer or between acquaintances on Facebook or in conversation with a stranger at a local restaurant. I then examine the moral sentiment that underlies the response and its maintenance via the circulation of images and phrases such as "a whale on the beach," an expression that Makah whalers use to refer to a successful whale hunt. The chapter concludes with an examination of what happened when the edibility of whales was challenged from the inside by a Makah elder, including the social pressures and normalizing mechanisms brought to bear in defending the orthodox point of view.

Not everyone in the Makah community "gets into the spiritual stuff," as one whaling captain put it, but many do. Chapter 3 examines the whale as a spiritual being. For most of the whalers with whom I spoke, the main point of their spiritual preparation was to safeguard against the inherent dangers of hunting a very large and powerful animal from a relatively small and fragile canoe. Others, however, felt that the preparation was necessary to make themselves worthy of the whale. They believe that if hunters prepare themselves properly, an animal's spirit will "offer itself" to them in order to fulfill its *telos* and feed the people with its flesh. It is difficult to generalize on this topic due to the normatively personal and idiosyncratic character of Makah spirituality, so this chapter recounts the spiritual views shared with me by individuals within and around the whaling effort, identifying recurrent themes. One prominent and widely shared Makah spiritual discourse is reflected in the refrain, "everything is connected."

In chapter 4, I explore an affinity that many Makahs recognize between local conceptions of nature and the utilitarian, quantitative ontology of the state. Complicating the anthropological view that modern resource management inevitably undermines the authority and legitimacy of traditional ecological knowledge (so-called), I examine how and why many Makahs feel they are able to merge their knowledge and mastery of the NMFS management system with more traditional spiritual discourses on whaling, effectively utilizing both registers in pursuit of their political aims. It is this overlap, I argue, that helped the tribe win federal approval for a whale hunt in the 1990s and that provides incentives for tribal officials to continue to work within the federal system.

Chapter 5 introduces the origins, motivations, and early tactics of the antiwhaling activists who have opposed Makah whaling since the 1990s. I focus in particular on one ad hoc, local antiwhaling organization called Peninsula Citizens for the Protection of Whales (PCPW). The Makah whaling conflict attracted sustained global interest and drew the attention of a bevy of environmental groups, but none played a more pivotal role than this small group led by Margaret and Chuck Owens of Joyce, Washington. The chapter begins by situating the group's goal of "speaking for the whales," within the broader history of antiwhaling sentiment in the United States. I trace the events that led to the failure of what I call interventionist activism, or the attempts by animal rights activists to interfere directly in the 1999 and 2000 Makah hunts. Although the story of antiwhaling activism in the Makah whaling conflict is one of adaptation to the authorized discourse of the state, the story begins with the activists' original defeat at the hands of a more conventional form of state power.

Following this defeat, antiwhaling activist groups involved in the Makah whaling conflict have been drawn deeply into the federal management system and its strict, exclusively quantitative enactment of gray whales. Chapter 6 examines this transformation and its consequences, beginning with the most significant development in the Makah whaling conflict in years: the publication of a study of gray whale population genetics that resulted in the restart of the court-ordered environmental impact statement on the proposed Makah hunt. Antiwhaling activists in the Makah whaling conflict now focus almost exclusively on "the science," as Margaret Owens refers to the policy-relevant studies that led to administrative delays. The strategic adoption of the fisheries management discourse has led to significant victories for the groups, as they have mobilized scientific uncertainty in bringing about indefinite delays to the Makah hunt; but activists' continued engagement with the state's interpretive framework comes at a cost, as it tacitly affirms a moral economy of stock-based management that excludes their preservationist aims in the long run.

ONE

It's Who We Are

Why whaling, and why now? If there is a consensus among scholars, it is that the continuing attempt to revive Makah whaling represents a reaffirmation of identity and sovereignty, or what the legal scholar Robert Miller (2000) calls "an exercise in cultural self-determination."[1] Walk down Neah Bay's main street or visit the collections at the Makah Cultural and Research Center (MCRC) and this interpretation will not seem farfetched. Whaling imagery is ubiquitous in the village. The tribe's insignia, an illustration of the primordial god Thunderbird clutching a gray whale, adorns the facades of businesses and government buildings.[2] A documentary on the revival of whaling produced by the Makah Whaling Commission runs on a continuous loop at the MCRC, where whaling is a centerpiece of the museum's permanent exhibit. Down on the waterfront, a giant, moss-covered whale skull sits in Keith Johnson's front yard not far from where his grandfather dragged his last whale ashore over a century ago.

"Who are the Makah? Whalers," Makah elder Greig Arnold put it succinctly. "That's the first word out of the mouth."[3]

In a 1998 op-ed for the *Seattle Times* titled "The Makah Manifesto," Johnson, president of the Makah Whaling Commission, wrote:

> No one can seriously question who we are; we are a small Native-American tribe whose members were the whalers of the American continent. We retain our whaling traditions today. It resonates through all of our people from the youngest to the oldest, and we don't take kindly to other people trying to tell us what our culture is or should be.

The assertive "we" of Johnson's manifesto leaves little space for dissent. Yet his own position in the community belies this notion of untroubled consen-

Figure 1.1 Whale skulls in the front yard of a home on the Makah Reservation
Photo by the author

sus. The head of the whaling commission and the former general manager of the tribe, Johnson is also a member of a prominent whaling family. Whaling was the sole occupation of elites prior to contact with Europeans, and the ability to identify direct "whaling ancestors" remains a salient distinction today. Those who can claim such ancestry tend to view whaling as an exclusive, hereditary right. "Every Makah has the *treaty right* to whale," Johnson told me on a number of occasions, always followed by a pregnant pause. To me, at least, his implication was clear: just having the treaty right did not mean one should climb into a whaling canoe and exercise that right. Wayne Johnson (no relation to Keith), the captain of the 1999 crew and never one to put too fine a point on things, once told a reporter, "I come from a whaling family, on my grandmother's side and my grandfather's side. It's in the blood. We have songs and dances around it. It kind of separates us from the rest. Not everyone in Neah Bay is a whaler."

The tribe conducted a household survey in 2002 that found an overwhelming majority of tribal members—over 95 percent—supported continued efforts by the tribal government to reinstate the whale hunt. There are clear indications that it was not always this way, however. At least a sizable minority initially opposed the resumption of whaling when it was first publicly discussed in the mid-1990s. According to two Makah elders who opposed the hunt, active support for whaling was limited to an influential minority, with a majority of tribal members expressing ambivalence or apathy for the project.[4] A former Makah Tribal Council member and ardent supporter of whaling recalls,

> In '96, '97, '98, alls our people saw was, oh the Council wants to do this so they can travel. That was a common, common grievance within the community. Aberdeen, Scotland. Monaco. Grenada. And, there's, um, fifteen, twenty people each time that went to these. And they were all related to the Council at the time. So, at first [whaling] was a real *divider*.

There is still a sense in the community that people from whaling families have more to gain from the resumption of whaling, while other, less traditionally powerful families may stand to benefit more from the modern political configuration of the tribe and its one-person-one-vote rule of governance.

As a pithy tagline for a complex set of motives, the claim that whaling is *who we are* ignores or conceals these unresolved tensions. The aphorism implies a monolithic and primordial Makah Tribe acting with a single will and purpose. None of this is out of the ordinary in the era of identity politics and race-based rights. Comaroff and Comaroff (2009) have shown how ethnically based political identities can lead to "thin" presentations of ethnic solidarity, purged of any nuance, that deny internal divisions while simultaneously reproducing internal inequalities (44).[5] Rather than simply stamp this sort of rhetoric with the imprimatur of the academy, we might ask how it is that whaling, once the sole privilege of elites, has come to be the activity most associated with what it means to be Makah.[6]

The timing of this shift offers some clues. The return to whaling gained broad support in the Makah community only after antiwhaling activists announced their opposition to it, coinciding with an increased emphasis on treaty rights in the tribal government's public discourse.[7] Among the respondents to the tribe's household whaling survey, nearly half cited the "maintenance of treaty rights" as the *primary* reason for their support of whaling.

Reflecting on the time he served in federal prison for his part in the 2007 rogue hunt, Andy Noel spelled out the stakes of the conflict for him:

> I have no regrets from doing that. I missed my family, but in the long run, I didn't give in. I *couldn't* give in. And, you know, it's much more than just us, there's a bigger picture and it involves Indians across the nation; I've always realized that, and I can say that I stood my ground. Not many people can say that. I've gone against one of the [most] powerful countries in the world and said, "eff you, man!"

As Andy's comments suggest, narratives that focus exclusively on what we might call the positive aspects of the hunt—the identity value added, if you will—miss the substantial meaning of the struggle for whaling rights to many people in the community. Narratives of cultural revival cannot account for the emphasis that is placed on the continuation of that struggle, especially for those who were not directly involved in whaling or who cannot claim to have had whaling ancestors. Much as Audra Simpson (2014) has shown how Kahnawà:ke Mohawk constitute their senses of self and sovereignty through various refusals (of Canadian passports, rights of settler-state citizenship, etc.), the Makah whaling conflict shows how identity can be constituted through refusals as well as through affirmation—in this case, a double negation: the refusal to accept *no* for an answer.[8] The refusal I speak of refers to tribal members' unwillingness to acquiesce in the face of resistance from antiwhaling activists, whose agendas are all too reminiscent of the infamous restrictions imposed on Indian fishing by the state of Washington throughout the early twentieth century.

Rather than *affirming* a static and unchanging Makah identity, the return to whaling was much more about the stakes of not backing down, of not giving up, of what would be lost if the tribe did not kill a whale once resistance was met. The community's response to antiwhaling activists played upon histories of colonialism and domination of which the activists themselves may not have been fully aware.

This dimension of the conflict is best understood by placing the struggle for whaling rights in historical context, beginning with the social importance of whaling prior to contact with Europeans and continuing through the so-called fish wars of the mid-twentieth century.

A Brief History of Makah Whaling

The organized pursuit of whales and other large marine mammals was a culturally salient and socially important feature of village life for the Kwih-dich-chuh-ahtx[9]—the Wakashan-speaking people of the semiautonomous villages near Cape Flattery whose ancestors would later comprise the Makah Tribe—in the several hundred years prior to contact with Europeans.[10] Like most Native American societies on the Northwest Coast, and especially those that historians and anthropologists have grouped under the Nootkan cultural type—or, as Arima and Hoover (2011) call them, the Whaling People (6, 16)—precontact Makah social organization was characterized by a high level of stratification based on hereditary status.[11] Prior to contact, extended family lineages formed the elemental unit of social organization, demarcated by cohabitation in one or more longhouses under a single chief or headman.[12] Each lineage was a microcosm of the social order and contained members of three distinct classes: an aristocratic class commonly referred to in ethnographic and historical literature as "chiefs"; a more numerous class of commoners; and slaves, members of neighboring indigenous groups who were either captured or purchased and were treated as property.[13]

As class-based occupations, whaling and other modes of resource gathering conferred relative prestige that reflected and reinforced the stratified social order.[14] Whaling, considered the most prestigious for its inherent danger and spiritual significance, was the hereditary privilege of the chief, an exclusive right he maintained by tightly controlling access to associated ritual knowledge.[15] It is unclear how important whaling was to the subsistence and economy of proto-Makah households in the centuries prior to contact. Some scholars have suggested that precontact Makah and Nootkan whaling was solely status driven: Drucker (1951) reasoned, for instance, that the social and ceremonial importance of whaling far exceeded its material value for the Makah's Nootkan relatives on Vancouver Island since "clearly the economic reward in proportion to the expenditures of time and energy was slight" (50). Swan (1870), on the other hand, speculated that whaling was once the most vital resource-gathering activity for Makahs, although he made his observations at a time when traditional Makah institutions had already been disrupted by colonialism. Colson (1953) gleaned that whaling

Figure 1.2 Makah harpooning a gray whale, ca. 1910
Photo by Asahel Curtis

was second only to halibut fishing in terms of material importance to the tribe. Waterman (1920) called halibut fishing the "mainstay of [the Makah's] existence" and stated that halibut was "more important from the point of view of their food economy than [was] the killing of whales" (9).[16] Based on the memory of their informants, early ethnographers reported that while blubber was prized, whale meat was not often utilized, either because it was not considered flavorful or because it spoiled before the whale could be butchered (meat spoils much more quickly than blubber, and it commonly took days to bring the dead quarry back to the beach; see Waterman 1920, 46). Even if only blubber is considered, available quantities were probably still massive enough to allow for consumption and trade in surplus oil.[17] Makahs traded extensively with other indigenous groups along the ocean coast and river basins.[18] Access to large quantities of whale oil was probably a factor in making the Makah "middlemen" in a vast trading network that extended hundreds of miles from Vancouver Island in the north to the Columbia River in the south.[19]

Trade with white settlers was an important factor in the consolidation of villages under nascent regional chieftainships after contact, and whaling may have played a key role.[20] In the spring of 1788, British trader John Meares anchored near Tatoosh Island and attempted to open trade with the Makah.[21] Meares had been drawn to the region by the prospect of a booming trade in sea otter pelts, the abundance of which had been reported by James Cook after his first encounter with Nootkans on Vancouver Island in 1778.[22] After sea otters were hunted nearly to extinction,[23] Makah traders turned to whale oil as the primary trade good.[24] The Hudson's Bay Company established a permanent presence in the region along with other fur trading companies, and Makah traders were eager to cultivate relationships with the company.[25] By the second quarter of the nineteenth century, the regional economy was dependent on systems of exchange between Native villages and European settlements.[26] Makah whale oil was sold to local settlers as well as to distant markets in Boston and Europe. By the 1840s the Makah were engaged in what amounted to large-scale commercial whaling, selling tens of thousands of gallons of whale oil to white traders and settlers each year.[27] In 1852 alone, the Makah landed twenty-four whales and sold an estimated sixty thousand gallons of oil on the commercial market.[28] As an indication of their market dominance, HBC officials at the time referred to whale oil as Cape Flattery oil.

In 1855, shortly following two calamitous smallpox outbreaks in the villages, Washington territorial governor Isaac Stevens brought his treaty council to Neah Bay to negotiate with the Makah. Stevens was in the midst of negotiating a series of treaties with Native bands in the region, the aim of which was to bring the groups under US control and eventually to effect their assimilation. Makah negotiators reportedly showed little interest in Stevens's offer of agricultural training and the usual treaty accoutrements, instead insisting on the guarantee of continued access to marine resources. The resultant Treaty of Neah Bay guaranteed the Makah Tribe the right of whaling and sealing in common with other citizens of the United States, the only Indian treaty to expressly do so. In a singular exception to the boilerplate Stevens treaty language guaranteeing tribes the right "to take fish at usual and accustomed places," Makah chiefs successfully prevailed upon Stevens to include an express mention of whaling and sealing.[29] The treaty included promises to outfit Makah whalers with materials and technology for whaling, reflecting the importance of the activity to the tribe and the region.[30]

Despite the pride of place given to whaling in the Treaty of Neah Bay, by the end of the 1850s the importance of whaling for the tribe began to decline. Successive smallpox outbreaks had interrupted the transmission of ritual knowledge that was so important to the practice, and the treaty's more egalitarian provisions eroded the traditional hierarchies from which whaling drew its social and cultural significance.[31] The Bureau of Indian Affairs (BIA) showed little faith in the whaling- and fishing-related promises of the treaty and instead pursued a standard agrarian assimilationist agenda, even though it was known at the time that the Makah Reservation contained very little land suitable for cultivation.[32] James Swan, a BIA agent who wrote the first ethnography of the tribe, was fascinated by local cultural traditions and lobbied for tribal fishing rights (his extensive diaries have been useful to contemporary Makahs in recuperating some lost traditions), but Swan's successors pursued more aggressive assimilation agendas and sought to stamp out Makah dances and potlatches.[33] In a poetic (and practical) display of resistance, the narration of which has become an oft-repeated contemporary trope, Makahs took the agricultural implements they were given and refashioned them into fishhooks and whaling lances.[34]

Commercial opportunities in seal hunting attracted the attention of young men away from the reservation in the years following the signing of the treaty, providing an alternative means to the wealth and prestige once exclusively afforded by whaling. After working as crewmen on white-owned sealing schooners in the 1880s, a number of Makahs earned enough money to purchase their own schooners; by 1892, tribal members owned ten sealing schooners, sometimes employing whites as navigators.[35] The profits from sealing allowed the tribe to consolidate their control over the Makah Reservation through the purchase of previously white-owned businesses in Neah Bay.[36] Makah whaling expeditions virtually ceased in the period between the 1860s and 1880s as sealing was booming.

Even if sealing was favored over whaling at the time, it is notable for the future significance of Makah whaling that it was the hunting of a marine mammal—the fur seal—that created wealth and sovereignty for the tribe at the same time that it provided the most effective form of resistance against BIA efforts at assimilation by drawing able-bodied young Makah men away from the reservation and the agency school.[37] Commercial sealing placed unsustainable pressure on seal populations, however, and by the 1890s seal numbers were in severe decline.[38] In one of the earliest conservation measures by the US federal government, severe restrictions were imposed on pelagic sealing, and in 1897, all pelagic sealing was banned, with the exception of Native hunts that used "traditional methods."[39] The manner of the restriction was likely as traumatic to tribal sealers as the restriction itself. In acquiring schooners and modern implements, Makahs had charted an apparent path to wealth and success in the face of modern challenges. In limiting them to only "traditional methods," the new law froze legitimate indigenous practice in a precolonial past, rendering Makah investments in modern technology useless.[40]

After the decline of commercial sealing and Makahs' subsequent failure to break into the commercial halibut fishery, Makah fishermen renewed an earlier petition to the government for help in acquiring industrial whaling ships, harpoon cannons, and other modern whaling implements.[41] But declining whale stocks made a commercial hunt impractical. By the turn of the twentieth century, the Makah were transitioning from an era of marine-based resource independence and wealth to an era of dependence and relative poverty. Makah whalers continued to harvest a few whales each year, using canoes, through the early 1900s, but by that time the hunts lacked much of their former social and ceremonial significance.[42] Calls for an international boycott arose in view of critically low gray whale numbers, and by the 1920s Makah whaling had ceased. According to local narratives, Makahs voluntarily stopped whaling in 1926 when they recognized that the gray whale population was in severe decline.[43]

The strength of tribal affiliation in the Pacific Northwest reached its nadir around the time that Makah whaling ceased. Prior to contact with Europeans, people living in indigenous villages in the Pacific Northwest had multiple modes of association available to them.[44] Extended family lineages provided the most basic unit of social organization and the primary basis of belonging, but individuals were also associated with winter villages, with summer resource-gathering sites, and occasionally with broader military and political alliances (although the last were usually ephemeral).[45] When Isaac Stevens negotiated federal treaties with Northwest Indians in the 1850s, he insisted on negotiating with "chiefs" and "tribes" even if it meant that he had to knowingly create them.[46] Tribal identities failed to immediately materialize following these negotiations, however, and Stevens had apparently given no consideration to the question of how to determine membership in the groups he had helped create.[47] Indians and settlers alike lacked consistent criteria for determining indigenous status, and in many cases they also lacked motivation to do so, with individuals continuing to claim multiple, overlapping affiliations.[48] Even immediate family members could have multiple, conflicting tribal affiliations.[49] The availability of wage work in hops farming and other industries drew Native peoples to racially diverse sites and urban centers where Pan-Indian identities emerged and blossomed.[50] In 1934, the Indian Reorganization Act created a ballast against this trend by tying federal benefits to tribal membership, but extended family membership remained the dominant organizing principle of Makah social life through the twentieth century, much as it still does today.[51]

Given the superordinate position of treaty rights in present-day discussions of whaling and in contemporary Native American political engagements with the United States more generally, it is striking to note that before the 1920s, treaty rights talk was far less prominent. People were undecided on whether they saw the treaty as a good thing, and they felt similarly unsure of whether their ancestors were worthy of admiration or scorn for their part in its negotiation.[52] Today, to claim descent from an acknowledged "treaty signer" is to claim distinction on par with whaling ancestry; treaty signers are now, as a rule, eulogized as wise and forward-thinking leaders. Interest groups like the Northwest Federation of American Indians (NFAI) played a key role in the transformation by focusing the bulk of their efforts on the issue of treaty rights—a rhetorical shift that was, according to historian Alexandra Harmon, far from inevitable. It proved a wise strategic choice, as ultimately the NFAI and other organizations were able to rally grassroots support around the treaties' unfulfilled promises. Treaty rights were effective as a Pan-Indian cause because the language of the treaties varied little between tribes. Moreover, laudatory narratives of treaty negotiations gave people a chance to redefine what had been perceived by many as a humiliating historical episode. Soon, tribal identities as such began to coalesce around complaints over sacred promises unfulfilled.[53]

Still, when Elizabeth Colson arrived in Neah Bay in 1941 to conduct her dissertation research in social anthropology, she observed a community that was in many ways indistinct from non-Native settlements in the region (1953, 50). Steady assimilation pressure and pragmatic parenting choices had begotten a generation of Makahs who, with few exceptions, received little overt teaching from parents and grandparents in the "traditional ways." Taken aback by Colson's interest in their past, Makah elders told her:

> We never thought the whites would ever be interested in anything we had so we tried to forget about it and learn the ways of the white man. Now that we have forgotten and live the way the whites do, people start to come around and ask us about what we had. . . . If I had ever thought that the whites would be interested in our doings, I would have listened to what the old people said. But we thought it was all gone and there was no use trying to learn the old ways any more. (126–27)

According to Colson, there was no pretense that practical knowledge about whaling had been passed on to the younger generation (176).

Visions of whaling pasts never fully disappeared, however. Colson reported that the memory of whaling maintained a special hold in the community and provided a sense of distinction. Whaling remained "an important part of the Makah imagination," according to Bowechop (2004, 409), "encoded in songs, dances, the communal sharing of food, and other social protocols."[54] Coté (2002) agrees that "whaling stories remained a central fixture in [Makah] oral traditions, passed down to the younger generation who only knew of whaling as something in their past that made them unique."[55]

Still, as late as the 1950s, Native Americans in the region remained largely dispersed, tribal governments lacked substantial authority, and tribal affiliation remained nebulous as both a political concept and a basis of personal identity.[56] A diverse set of policies and proposals grouped under the heading of "termination" (referring to the proposed end to federal trust status for Indian tribes) guided national political debates about Native Americans through much of the 1940s and 1950s.

The Fish Wars

All of this changed beginning in the 1950s as the struggle for fishing rights provided a crucial turning point for Native American tribal identities in the Northwest, becoming the crucible in which Northwest tribes were remade into the configurations recognizable today. Although the seeds of discontent had been planted earlier, the struggle reached the height of its intensity between 1950 and 1974, during a period marked by almost a quarter century of protest, litigation, and struggle. The "fish wars," as they are commonly known, culminated in dramatic victories for Washington tribes and established the patterns that would later give Makah whaling much of its shape and political significance.

The state of Washington had sought to undermine the Treaty of Neah Bay almost immediately after it had come into being, first by arguing that its very existence abrogated tribal fishing rights (Sepez 2001, 60) and later by cracking down on off-reservation fishing under the label of conservation (65–66; see also Reid 2015, 239–42). Tribal members still fished outside of the reservations, but they did so secretly, often at night in order to avoid arrest, and those who did were often branded as poachers by the state. The Washington State Game Department reportedly produced a propaganda film entitled *The Problem of Indian Fishing*, which portrayed Indians as dangerous "predators" who would destroy the steelhead population if left to their devices.[57] Legal disputes involving Indian tribes during this time almost always led to debates about hunting and fishing. Native American witnesses appearing before Congress for the aforementioned NFAI lawsuit used the occasion as an opportunity to decry their marginalization from their traditional hunting and fishing grounds.[58]

In 1951, Makah fishermen challenged the state of Washington's conservation-based prohibition against net fishing in the Hoko River, a salmon and steelhead stream that lies roughly fifteen miles east of the Makah Reservation.[59] The court ruled in favor of the Makah, concluding that Hoko River salmon were an essential component of the tribe's food supply and that the state "had not sustained the burden of proving that such regulation was necessary for conservation." The Makah lawsuit became a cause célèbre and, according to Harmon (1998), sparked "a twenty-year spiral of action and reaction" between treaty tribes and the state of Washington, set against the backdrop of the broader American civil rights movement. The result was spontaneous civil disobedience of net-fishing bans across the state.[60] Members of the Puyallup and Nisqually tribes staged high-profile "fish-ins," sometimes joined in protest by celebrities like Marlon Brando.[61] In his memoir, longtime Makah tribal attorney Al Ziontz recounts his initial conversations with a Makah official after the tribe had first expressed interest in retaining the services of Ziontz's law firm:

> "I'll tell you why I'm calling" [the official said]. "You know I've always felt the Makahs were getting a raw deal. Their treaty rights mean nothing to the state. A lot of us younger Indians around the country feel it's time to kick ass.... We're gonna demonstrate on the rivers and let them arrest us, like they always do. But this time we're going to try to have some celebrities with us.... Marlon Brando is coming and we're going to have a fish-in, just like the black people had sit-ins." (Ziontz 2009, 4, 50)

The fish-ins brought increased visibility to the tribes' struggle for fishing rights. Billy Frank Jr., a Nisqually Indian, rose to prominence during this time for his civil disobedience of net bans. Frank would remain an influential figure in Pan-Indian activism and later as a lobbyist for the collective interests of Washington treaty tribes.

Continued unrest and treaty rights litigation culminated in *United States v. Washington* (1974), a seminal court case argued in the Western District Court of Washington that is now known locally as the Boldt decision. Judge George Boldt ruled that the Stevens treaties entitled Native fishermen to one-half of the annual commercial take of all harvestable salmon that passed through each tribe's "usual and accustomed fishing grounds," or the lands and waters traditionally held and exploited by each respective tribe.[62] The

Figure 1.3 Marlon Brando and Billy Frank
Courtesy of the Museum of History and Industry, Seattle

decision relied on the "reserved rights" doctrine established by the Supreme Court in *United States v. Winans* (1905), which holds that rights not specifically granted to the United States in treaties are maintained by the tribes. In his decision, Boldt expressly quoted a key passage from *Winans* that Indian treaties represented "not a grant of rights to the Indians, but a grant of rights from them."[63]

The Boldt decision signaled a major advance in the struggle for indigenous resource rights, and it contributed to an emerging sense of tribal self-determination and sovereignty among Northwest coastal peoples. This feeling was compounded for the Makah by a serendipitous archaeological discovery in July 1970 at Ozette, the site of a precontact village that had been abandoned in the nineteenth century. Hikers found that winter storms had unearthed part of a Makah whaling village that had been buried by a massive mudslide almost three hundred years earlier.[64] Informally dubbed the Pompeii of the Americas, the discovery of numerous well-preserved whale remains, whaling implements, and whaling-related artwork at the site provided tangible evidence of the Ozette villagers' whaling past.[65] Dozens of young Makahs were recruited to help with the excavation, sparking interest in Makah traditions and quickening calls for cultural revitalization.[66]

Whispers of Whaling

It was in this atmosphere of cultural revitalization and resurgent resource sovereignty that Makah commercial fishermen first began to talk seriously about a return to whaling.

In the salmon season of 1979, a gray whale became entangled and drowned in the marine set-net of a respected tribal fisherman. Although the catch was inadvertent, federal authorities confiscated the whale and threatened to assess the fisherman a steep fine for violating the Marine Mammal Protection Act (MMPA). Accidental entanglement was not an uncommon occurrence at that time,[67] and in such cases fishermen sometimes cut the whale free and furtively sank it to avoid fines. But this time was different. The Makah Cultural and Research Center, built to house artifacts from the Ozette archaeological dig but already emerging as a site for representing and performing Makah identity, had opened its doors earlier that year.[68] The Boldt decision had been upheld by the US Ninth Circuit Court of Appeals. Two archaeologists from the Ozette dig had recently published a book called *Hunters of the Whale* that spoke in detail of the Makah Tribe's proud whaling past.[69] Emboldened by their legal victories and angered over this latest encroachment by federal officials, Makah fishermen drafted a letter to the Makah Tribal Council asking it to take active steps to defend fishermen's treaty right to take marine mammals. The fishermen felt they should be able not only to keep without penalty the whales they inadvertently caught in their nets but also to hunt whales openly.

During this time, federal officials were still considering the question of how rigorously to enforce the restrictions of the MMPA as it applied to tribal sealing. In 1981, federal law enforcement officials stormed the home of Greg Colfax, a renowned Makah carver and educator, and seized seal meat from his freezer and smokehouse. I first met Greg when I stayed in his modest hotel, two adjacent rooms that he had recently refinished above Linda's Woodfired Kitchen, the pizzeria his wife ran, which offered the best food in Clallam County by a fairly wide margin. I can practically hear the wry amusement in his voice in the description of the raid that he provided to Jennifer Sepez in 2001:

> That evening at dinnertime my wife and family were sitting around the dinner table, and next thing, I look outside, and I'm surrounded. There's tribal

Figure 1.4 Wayne Johnson

police all around, there's *federales* around, there's NOAA [National Oceanic and Atmospheric Administration], all the representatives. And pretty soon: knock, knock, knock. I answer the door. Badge. "Got a search warrant. We're here to pick up materials, the seals." I says, "You lousy guys." So, they came in. . . . They wanted to have the seal meat. I says, "You don't have a right. That's my food."

News of the raid spread quickly across the community, and several weeks later a "sealing party" was held, a public display of resistance where "the guest of honor was a whole pile of seals."[70] Additional raids led many to fear they would be the next target. For some, it was all too reminiscent of government tactics during the early stages of the fishing controversy decades earlier.

In 1986, two elder fishermen again approached the tribal council, this time with two requests: "We want to change our name back to Kwih-dich-chuh-ahtx," they said, "and we want our marine mammals back."

Working with the Feds

The majority of academic and popular accounts of contemporary Makah whaling begin with the gray whale "coming off" the Endangered Species List in 1994, as if it happened spontaneously and the Makah, marveling at the serendipity of it all, reacted by announcing their intent to hunt gray whales again.[71] In fact, Makah leaders played a central role in petitioning for the "de-listing" of the gray whale, using existing relationships with federal officials to lobby for the species' removal.[72]

The decision to seek approval to hunt gray whales, as opposed to other species, was a considered choice. Archaeological data and oral histories both suggested that gray and humpback whales were exploited at similar frequencies prior to contact,[73] and today most Makah whalers will say that humpback whale meat is far superior to gray whale meat in terms of taste and consistency. Makah fishermen perceived at the time that gray whale numbers were on the rebound, however, and a long history of IWC-approved indigenous gray whale hunting in the Russian Arctic meant that international regulators had an established procedure for calculating sustainable "removals" on the stock. According to Dave Sones, a Makah commercial fisherman and an early proponent of whaling, this made the gray whale a "good fit" for the tribe's plans.

From the beginning, tribal officials disagreed over whether commercial whaling could be a viable option for the tribe. Sones told me that although some people said a single gray whale could fetch half a million dollars in Asian markets, the commercial option was never viable for the Makah due to domestic political pressures. Still, he felt it was important to reserve the commercial right for future generations. Others were more optimistic about the commercial prospects of whaling, viewing the acquisition of fishing and sealing rights as propitious signs in the march toward tribal sovereignty.[74] Micah McCarty recalled walking into the office of Dan Greene (a young Makah fishermen and tribal council member who would later serve as a nominal co-captain of the 1999 hunt) around the time of the delisting: "On one of the drawing boards they had a whole list of sea mammals, marine mammals, and the gray whales were on the top of the list." Greene asked Micah's father, John McCarty, "How would you like to go whaling?" (Coté 2010, 132). At least one member of the tribal council envisioned the construction of a marine mammal processing plant in Neah Bay that would process multiple species of whales, porpoises, seals, and sea lions for sale in overseas markets.[75]

Makah leaders were hesitant to go forward with a hunt on an officially endangered species. "The Alaskan Natives, they do hunt on a listed species," Sones told me over the phone one day from the cabin of his commercial fishing boat, "but they've been working with it for a long time and they got a lot of research and science invested there. We thought it would be an easier process to establish our traditional hunting through a species that was being delisted." In addition to presenting fewer bureaucratic headaches, delisting was also in the tribe's interest, Sones said, because it would help establish that the Endangered Species Act was a "two-way street," meaning that successful recoveries could result in the removal of animals from the list. Makah officials asked the Northwest Indian Fish Commission (NWIFC), a lobbying organization created in 1975 to support the natural resource management activities of the twenty "treaty tribes" in western Washington, to send a letter to the Department of Commerce urging the removal of the Eastern North Pacific gray whale from the Endangered Species List. The resulting letter, penned in March 1991 by NWIFC Chairman Bill Frank Jr. (of 1960s "fish-in" fame), utilized whale population survey data provided by National Oceanic and Atmospheric Administration (NOAA) biologists to argue that gray whale stocks had satisfactorily rebounded from their postindustrial whaling lows. When the National Marine Fisheries Service (NMFS) failed to publish the proposed change by the end of summer as they had promised, the NWIFC grew impatient. In October, Frank sent a strongly worded letter to the Commerce Secretary requesting that he "personally investigate the reasons for . . . delays and direct NMFS to actively pursue the immediate de-listing of the Gray Whale under the Endangered Species Act."[76] A representative of the Commerce Secretary replied one month later, informing Frank and the NWIFC that the proposed rule to delist the gray whale had been published in the *Federal Register* and was awaiting public review and comment.[77] During the review period, John Twiss, the executive director of the Marine Mammal Commission (an independent US agency charged with overseeing marine mammal conservation policies) sent a sixteen-page letter to NMFS recommending that in light of continued threats to gray whales posed by human activities, the species should be downgraded to "threatened" status rather than removed from the Endangered Species List entirely.[78] NMFS rejected the recommendation and moved forward with the delisting process. William Fox, Assistant Administrator for Fisheries at NMFS, wrote to Twiss in reply that the Marine Mammal Commission's concerns seemed to be based on anticipation of "future events" and that the increasing abundance of the stock suggested Eastern North Pacific gray whales have "the resiliency to adjust to human activities with little apparent adverse effects."[79]

The success of the petition to delist the Eastern North Pacific gray whale was an indication of the Makah Tribe's vastly improving relationships with federal fisheries management agencies. Just a decade earlier, clashes with federal law enforcement agents had been the spark for the tribe's aggressive pursuit of marine mammal harvest rights. NMFS scientists were now assisting and encouraging the tribe, and there was talk even of federal grants and other financial support to help the Makah revive their whaling tradition, which the tribe would eventually receive in the amount of $335,000.[80]

When two elder fishermen first approached the tribal council to talk about getting their marine mammals back, the request was about more than gaining access to the resource. Makah hunters had been surreptitiously shooting, transporting, and butchering seals for decades, and those practices continued despite a few high-profile raids. Access was not the issue. The fishermen's request, then, was not about regaining the ability to hunt marine mammals but about the state's recognition of their right to do so.

Aberdeen and Its Aftermath

The tribe kept their whaling plans hidden from the public until after the gray whale was officially removed from the Endangered Species List in 1994. Well after the delisting process was in motion, Sones, in his capacity as director of Makah Fisheries, told reporters that the tribe only wished to see the gray whale downgraded from "endangered" to "threatened" status and had no intention to hunt gray whales.[81] He said the tribe's intent behind the petition was to free up federal money for monitoring other endangered species of concern to the tribe. Why Sones chose this public misdirection is unclear, but soon after the delisting was officially approved, the chairman of the Makah Tribal Council wrote to the Department of Commerce requesting technical assistance in developing a management plan for the harvest of five gray whales per year (one whale for each "ancestral village").[82] The request was not a complete non sequitur—NMFS had already approved a Makah request the previous year to harvest fifteen harbor seals, of which four seals were eventually taken.[83]

At the behest of federal officials, the tribal council agreed to first seek approval for a whale hunt from the International Whaling Commission (IWC). They agreed to seek this permission as a courtesy, stating clearly that they did not feel IWC approval was necessary for them to exercise their treaty right.[84] The Makah request put the United States in a delicate position at the IWC meetings. Although the American delegation had for some time supported the continued hunting of endangered bowhead whales by Alaskan Inuit whalers, the United States had otherwise built their platform around a zero-tolerance policy for new whaling ventures globally. The Makah Tribe's long hiatus from whaling also presented a potential obstacle. Every other indigenous group that hunts large cetaceans in North America (i.e., Alaskan Inuit groups) does so under an IWC exemption for subsistence hunting, but because the Makah had not whaled in at least seventy years, it was questionable whether they could claim such a need under IWC rules.

Tribal officials called upon Ann Renker, a cultural anthropologist and co-author of the "Makah" entry in the Smithsonian *Handbook of North American Indians*, to address these issues in the tribe's "needs statement," which the IWC requires for all aboriginal subsistence whaling applications. In the statement, Renker introduced the concept of a "cultural need," stating that the reintroduction of the whale hunt would "provide physical subsistence for the bodies of Makah people, and *spiritual subsistence* for its traditions and cultural identity."[85] In later versions of the statement, Renker would update this passage to read, "spiritual subsistence to the soul of the Makah people."[86] The statement suggested that whaling could ameliorate a range of issues plaguing the community, from diabetes and shortened life expectancy to the social and behavioral problems affecting Makah youth (1997, 23–24). These claims were similar to those made by Keith Johnson in his "Makah Manifesto," in which he wrote, "Many of our tribal members feel that our health problems result from the loss of our traditional seafood and sea-mammal diet. . . . We also believe that the problems that are troubling our young people stem from lack of discipline and pride and we hope that the restoration of whaling will help to restore that."[87]

At the IWC meetings in Aberdeen, Scotland, commissioners from France and the Netherlands questioned the existence of a "continuing tradition" for the Makah given the tribe's long hiatus from whaling.[88] Renker countered this suggestion in her testimony by comparing the Makah hiatus from whaling to the plight of European Jewish culture during the Holocaust. She argued that the traumatic interruption of a tradition did not mean the tradition could or should be abandoned. But the final blow to the Makah request came when a contingent of Makah elders who opposed the hunt spoke out against it. The United States was compelled to postpone the Makah request until the following year.

After the IWC meetings, public backlash to the hunt from outside the Makah community intensified. The Sea Shepherd Conservation Society sent a memo to its supporters outlining the dangers of a Makah hunt for global whaling precedent. Protesters held rallies in Seattle and other nearby towns. Some of the protesters' discourse played on bitterness stemming from the special privileges awarded to Native fishermen in the Boldt decision.[89]

Following this public backlash, tribal leaders altered their plans: they would carry out whale hunts purely for "ceremonial and subsistence" rather than commercial purposes.[90] Renker began preparing a strategically worded "household whaling survey" to produce data that would suggest unanimity among Makah tribal members with regard to their attitudes toward whaling. The proposed hunting methods were also revised. Some early proponents had favored large commercial vessels with bow-mounted harpoons, but influential voices within the newly formed Makah Whaling Commission called for a canoe-based hunt and a handheld harpoon. The Coast Guard preferred a larger, motorized boat to ensure the safety of the crew, but after more than six months of negotiation the two sides settled on a compromise: the whale would be harpooned from the canoe and a shooter would deliver the killing blow from a motorized chase boat. To conform to IWC rules requiring the "most humane kill" possible, the tribe contracted a veterinarian from Maryland who had previously consulted with Inuit whalers on killing methods. The veterinarian procured and tested a special .50-caliber whaling rifle for the Makah to use in the hunt.[91]

In 1997, the US delegation was successful in securing IWC permission for the Makah hunt. The manner of the approval was controversial. The IWC amended its schedule of rules to include Renker's concept of cultural need under the aboriginal subsistence exemption, but the matter never went before the full IWC committee again.[92] Instead, the US delegation secured an annual quota of five gray whales for the Makah through an informal swap with the Russian delegation wherein Russian Chukotka Natives received an annual share of the Alaskan Inuit bowhead whale quota.[93]

The tribe, concerned that they were losing the public relations battle and increasingly believing that such things mattered to their cause, hired former attorney Al Ziontz to serve as their manager of public relations. Ziontz immediately suggested a press conference at the Makah Cultural and Research Center.[94] They also invited a documentary film crew from Germany to film parts of the preparation and execution of the hunt.[95]

As the focus turned to preparation for the hunt, multiple paths to whaling-based identity became available for members of the tribe. An immense administrative apparatus grew up alongside the Makah Whaling Commission to interface with administrative bodies at the federal and international levels. An outgrowth of the tribe's Department of Fisheries, it provided additional opportunities for people to become directly involved with whaling. Substantial turnover in the would-be whaling crew also meant that dozens' of Makah men would experience at least some preparation and training for a hunt.

Patterns of Struggle

The decision to seek approval from NMFS and the IWC angered some members of the tribe. John McCarty, the first executive director of the Makah Whaling Commission, felt the tribal government should not be compelled to negotiate for something that was protected by their treaty.[96] Several members of the tribe likewise told me that working with the federal government was a mistake because it sent the message that treaty rights were negotiable.

These disagreements tended to break along generational lines. The fish wars taught the elder generation of Makahs a fairly simple model of protest: go first and fight later.[97] Those who opposed seeking IWC approval had, by virtue of their age or employment history, the most direct experience of the Northwest tribes' battles with the state of Washington for resource rights. Jerry Lucas embodied this ethos one day over morning coffee when Wayne Johnson, the captain of the 1999 whaling crew, asked Lucas why he always "fought him" over the whaling.

"I *wouldn't* have fought you if you'd just gone and done it!" Lucas boomed in response. "But you had to go ask for permission!" Later, in an interview, Lucas expanded on his reasoning:

> Going whaling, my opinion at the time when I was on the tribal council, I said we don't advertise it, it's our treaty right, I said, let's go do it. And we don't need to ask anybody. There was a couple members that were up and coming on the tribal council, younger than me, that [didn't mind] the publicity. [But] there's some things you don't flaunt in the spotlight and I believe that this portion of our treaty we don't flaunt, we just go do it. . . . I said I don't want anybody else involved, just the Makahs. And I liken this, my attitude, come this way 'cause of the old Indian named Sohappy from Yakima, that got arrested for fishing in the Columbia, and I said he exercised his treaty right, yes he wound up in jail, he wound up in court, but . . . it come to exposure some of the rights that treaty tribes have.

Lucas attributes the differences in opinion about seeking IWC approval to an increased comfort with "the spotlight" among his younger colleagues. Those differences probably have as much to do with the experience of the younger generation of fishermen after the successes of the Boldt decision. The young fishermen believed that meaningful change could be won within the system. Members of the old guard like Lucas, on the other hand, were turned off by the administrative "hoops" the tribe was compelled to jump through. The 2007 rogue hunt more closely resembles Lucas's preferred pattern of struggle.[98] The rogue hunt was generally considered a failure, however, because it was unsuccessful as a protest: it damaged, rather than bolstered, efforts to reacquire the legal right to hunt whales (I will further explore the details of the rogue hunt and the circumstances of its failure in chapter 7).

Uniting for Treaty Rights

Not everyone in the tribe was privy to early discussions about a return to whaling, and when plans for a whale hunt were first announced to the community in the mid-1990s, not everyone was supportive. In fact, as much as half the tribe opposed the idea of a whale hunt. Some people in the community thought whaling was a slush fund for tribal council members, a waste of money. They thought the political and financial risks were too great, that a hunt might endanger future grants and jeopardize federal support for other projects. As the former tribal council member I quoted at the beginning of the chapter stated:

> At first it [whaling] was a real *divider*. And it stayed that way for a long time, until Paul Watson came to Neah Bay. And until animal welfare terrorists came to Neah Bay and said "No, we're not gonna allow this." Then you saw Makah come together.

The hunt gained broad support among the members of the Makah Tribe only after activists announced their intent to oppose the hunt. When they encountered resistance, the tribe figuratively closed ranks. A minority of Makahs still maintained that whaling wasn't worth the backlash it would cause, and they argued that abandoning plans to hunt whales would be the politically and fiscally prudent thing to do. But their voices were drowned out—if not immediately, then by the indignant chorus that greeted external resistance to the tribe's plans.

Had antiwhaling organizations not intervened—had Sea Shepherd not brazenly anchored their repurposed Coast Guard cutter in the Makah harbor for months in protest of the proposed hunt—it is possible that events might have turned out differently. Internal divisions and decades-old tensions between Makah families might have prevailed and snuffed out the kind of fierce unity needed to push past the opposition. But activists had unwittingly placed in the Makah's proverbial backyard a tangible symbol of everything tribal members had struggled against for decades in the battle for resource rights, a struggle that had served as a crucible for emergent tribal identities.

Talking about whaling with tribal members today, one notices that the struggle for treaty rights is usually emphasized over goods internal to the practice itself. The majority of people I spoke with told me that their primary motive in seeking to reinstate tribal whaling was to "protect the treaty right." Several described whaling as a "use-it-or-lose-it" proposition.

"It's not so much the whaling; we are securing the treaty right," tribal council member David Lawrence told a *Seattle Times* reporter in 2002.[99]

"Now that we've established the treaty right," another tribal member told me, "I don't care if we ever go again."[100]

"If the tribe didn't fight for whaling," a Makah administrator who played a critical role in the whaling effort declared to me in an interview, "I wouldn't be able to call myself Makah."

In my third of four extensive interviews with Greg Colfax, a local carver and cultural leader, I asked about the moment the whale was brought back to the beach in 1999. "How did you experience that?" I asked him, expecting the kind of earnest but apolitical philosophical reflection I'd grown accustomed to hearing from him. Instead, he replied:

> I think that it was a statement to the world that we will not go away and that the reason we are here is because our grandparents and our great-grandparents want us here and want us to hold on to this place. . . . It was our reply to everything that was done to my family, that I know about, everything that was done to my dad, and to his dad, and to all the members in my family, in their struggle to stay here. Their struggle against the Indian agent, their struggle against real estate people, their struggles, when they go to town, the lack of respect they received, y'know? That was a statement to all of that, and it said we are, we have some pride. We worked, and we succeeded.

Hearing this, I shared with him an early version of the interpretation I have presented here, that the whale hunt was as much about the battle with outside opposition as anything else.

"I think that part of the statement is true," he replied, "because the opposition was so enormous. I would have to agree with you to a certain extent that my acceptance of the—of the hunting of the whale and in that moment of bringing the whale to shore . . . [was] a statement in opposition to those who are against us, and who were against my family for years and years and years and who provided the impetus of the oppression against Makah people."

According to all available accounts, this was the prevailing sentiment in the celebratory potlatch following the 1999 hunt.[101] Al Ziontz, the tribe's longtime attorney and director of public relations for the 1999 whale hunt, summed this up when he spoke before those assembled there:

> You Makah people have a right to be proud today. Once again you have won back your sacred rights. Many of you remember the *fight* to save Ozette from those who wanted to sweep it into the Olympic National Park. You won that fight and Ozette is Makah today, along with everything that came out of the ground there and now sits in the Makah museum. You won back Tatoosh and Waadah, your ancient lands, which the federal government tried to take from you. You won the right to fish in the Hoko, and with the Boldt decision, you have won the right to use your treaty rights in all Makah usual and accustomed places, on the rivers and on the ocean. And now today, you celebrate the treaty right your ancestors preserved for you: the right to hunt the whale. You faced powerful enemies in this, and the fight may not be over, but I know that in the end, the Makahs will win.[102]

When the whale was brought to the beach after the successful 1999 hunt, the Makah canoe was joined by a delegation of canoes from other coastal tribes, some of whom had publicly expressed ambivalence about the hunt but came to express solidarity in the struggle for treaty rights.

Unresolved Tensions

Outside opposition galvanized support for Makah whaling across the community's many families and factions, and today few Makahs would publicly question the hunt. The echoes of a stratified social order have not fully disappeared, however, and tense questions persist over who should be allowed to whale and who should not.

Over the last 150 years, a number of powerful forces have encouraged the production and adoption of a monolithic Makah tribal identity: the Treaty of Neah Bay and its institution of the one man–one vote rule; systematic efforts by Indian agents to eradicate traditional social organization;

the indigenous reuptake of ethnographic representation that does not distinguish between family lineages; legal classifications and benefits stemming from US Indian policy reforms in the 1930s; identity-based politics and the struggle for resource rights; and the expectations—based on ossified concepts of culture—of organizations like the International Whaling Commission. To this we might also add the political investment of tribal councilmen, administrators, and other powerful actors in the notion of "the tribe" as a governmental body. Yet despite all of this, whaling remains deeply associated with claims to elite social status and hereditary privilege.

As I mentioned earlier, the extended family—not the village or the tribe—was the elemental unit of social organization prior to contact with Europeans. Kin-based hierarchies began to lose their influence in formal government as early as the 1850s, and the election of the first Makah Tribal Council in 1877 signaled the end of the formal political significance of the traditional status system.[103] But the primacy of the family persisted in the domain of intellectual property such as ceremonial dances, technical knowledge, rituals, and other practices later delineated as "cultural" by members of the community. This is reflected in the archiving practices of the Makah Cultural and Research Center, where the unknown family affiliation of the longhouses uncovered at the Ozette archaeological site poses challenges regarding who among the tribe should be allowed to see and handle artifacts recovered from them.[104]

Today, membership in extended families remains the dominant organizing principle of informal social life in the Makah community. Very little socialization in the private sphere occurs between members of different extended families. Individuals identify strongly as members of one family lineage or another, even if it means emphasizing some ties while de-emphasizing others. Different families maintain different levels of investment in the notion of a single Makah tribe. For some members of the whaling crew, in fact, the tribal hunt was the first time in their lives they had ever consciously done anything significant for the tribe per se rather than for their family.

Perhaps most telling is how collective memory continues to operate along family lines, as in a discussion I had with Keith Johnson about when the tribe had hunted its last whale.

"My grandpa played on the tail of his father's whale, in 1907," Johnson said, "in front of [what is] my house now. He played on the tail of the last whale in. Here I thought the last whale was down at the other house. . . ."

"Nobody can agree whether it was 1907 or 1930," I replied naïvely.

"*Our* family's last whale in. No one argues with that one. 1907. There

It's Who We Are / 57

Figure 1.5 Wilson Parker posing with harpoon and float, 1915
Photo by Edward Curtis

was other whales probably come ashore. Probably a lot of other things. But *ours* was 1907."

Another elder once explained to me that as a member of the community, "you aren't interested in other families' business. You don't even ask about it." Keith's emphasis on *his* family's last whale exemplifies this, as does his gloss of everything outside his family's history as *other things*. Those other things might as well have been happening in another village.[105]

A controversy surrounding an iconic photograph taken by Edward Curtis around the turn of the twentieth century illustrates some of the salient tensions surrounding claims to whaling pasts. The photo shows a Makah man, Wilson Parker, posing with a whaling harpoon and float. The picture is prominently displayed in the Makah Cultural and Research Center and has been reprinted widely in academic literature. Descendants of the man in the picture have taken a central role in contemporary whaling, and they consider the photo to be evidence of their family's elite traditional status. But one whaler from a rival family who was beginning to feel excluded from the contemporary whaling process once said to me, exasperated: "The guy comes out of nowhere, poses for one picture, and now he's a whaler?! Now his family are whalers?!"

The tribe's solution to the internal strife flowing from the legacy of hereditary privilege was a tribal hunt in 1999, with representatives from different families comprising the crew. Although this arrangement was intended to cast the hunt as a tribal venture, most of the whalers I spoke with framed their participation in the hunt as, first and foremost, a matter of representing their family on the crew. Some tribal officials had hoped that the 1999 hunt would serve as a kind of harbinger of a new cultural dispensation, after which *everyone* in the tribe would be "whalers." Clearly skeptical of this conceit, Wayne Johnson reflected on the state of things one day as I watched him fillet sockeye salmon on the banks of the Wa'atch River. He believed the tribal council was trying to get him and a close associate removed from the Makah Whaling Commission. "It's almost like my slaves are getting back at me. But I don't like to say that since this is the twenty-first century. And after '99, we're all whalers."

The tribal hunt was followed in 2000 by "family hunts," with permits issued to "whaling families" upon request. As the process moves forward and the tribe seeks to maintain its IWC-allocated quota of five gray whales per year, there is continued conflict over which families control the process and questions about who should attend the IWC meetings to represent the tribe. This unresolved tension continues to run through all discussion of whaling in the community.

Whaling and Makah Identities

The identities that are made and remade in the Makah whaling conflict are complex and multiple, as reflected in the polysemy of the *we* in the Makah aphorism "Whaling is who we are." This chapter has focused mainly on two of the identities contained in the subject position of that phrase. The first is a pugilistic collective self, a Makah Tribe that fights for treaty rights against the onslaught of colonialism and threats of dispossession. The figure of the "treaty signer" serves as the exemplar for this version of Makah identity, and it is telling that historically speaking, treaty signers were not held in such esteem until the emergence of tribal identities as such in the mid-twentieth century.

Scholars have observed that identities are built on exclusion, acquiring their meaning largely in contrast to what they are not.[106] A Makah identity based on the concerted, tenacious struggle for resource rights risks alienating only those who would relent in the face of outside pressure. This counterpart is a fiction, albeit a historical one. Few living persons are excluded, making this version of Makah identity much more ecumenical in its appeal. The identity of the whaler who possesses hereditary rights, on the other hand, takes its meaning from a contrast with others in the community, with nonwhalers. When a Makah whaler states that whaling is in his blood or in his DNA (as Keith Johnson once told me), we might read it as a conspicuous naturalization of race or tribal membership, but clearly it is more than that.[107] Such statements also assert kinship-based privilege. Wayne Johnson and Theron Parker refer to their blood or DNA not just as members of the Makah Tribe but also as heirs to whaling families.[108]

These two concurrent identity claims have a better chance of hanging together when they are galvanized by outside opposition, such as the fierce resistance Makah whalers encountered from animal rights activists in the 1990s. When these activists occupied Makah marine space and hurled insults over loudspeakers from a foreboding black ship parked in Neah Bay's harbor, when they wrote letters to the editors of local newspapers that revisited colonial atrocities for rhetorical effect, and when they opposed the tribe in domestic and international regulatory forums, they had the effect of uniting the Makah community behind memories of oppression and future dreams of resource sovereignty. As memories of the direct, physical clash between whalers and antiwhaling activists become more distant, we may expect the tribe will have more difficulty suppressing internal tensions and establishing consensus among its members.

As I hope this chapter has shown, the identities risked and forged in the Makah whaling conflict are not just multiple but also contingent. As Patchen Markell (2003) has observed, discourses on identity within a contemporary politics of recognition start with a misrecognition, as they tend to overstate what we might call the sovereign performativity of the group or individual and their ability to dictate their own biography (and project that biography into the future; see also Londoño-Sulkin 2012, 21). In the present case, the claim that "whaling is who we are"—whatever its pragmatic intent—is, thus, more accurately rendered as "whaling is who we have been, and whaling is who we would like to be," the realization of which depends, in part, upon actions and circumstances that lie inexorably beyond the control of the subject who utters such aspirations. The serendipitous discovery of Ozette, the legal disposition of its artifacts, and the temporal intersection of the discovery with the American civil rights movement come to mind as potent examples of just such contingency.[109] Other events include the judicial successes and commercial aspirations that opened up whaling as a pathway for traditional whaling families to (re-)assert traditional status, the recovery of gray whale populations, and the activist opposition that made a unified response into an existential struggle for tribal sovereignty. To name just a few.

Because contingency has the ring of happenstance, and because it is commonly opposed to the *essential* and the *necessary*, the Makah Tribe and its spokespersons at the Makah Cultural and Research Center have perceived it to be in their interest to conceal this contingency as much as possible, from their administration of the household whaling survey to their influence on outside academic work conducted in and around the Makah Reservation. In the context of a politics of recognition within a colonial history and present, they seek to present whaling as the inevitable expression of an essential, unitary Makah-ness; an affirmation—whaling is who we are.

TWO

We Eat Them

Ask anyone in Neah Bay what they think of gray whales, and there is a good chance you will hear some variation on the same single-word reply: "Yum!"

"Whale, the Makah word for dinner," commented one tribal member in an online post shortly after Micah McCarty, the once-and-future chairman of the Makah Tribe, posted an aerial photograph of a gray whale taken by a local news organization to his Facebook page, along with the caption "Yummy!"

"Are you bringing home dinner?" a commenter quickly asked.

A political cartoon I saw hanging on a refrigerator in Neah Bay (figure 2.1) compares the English term *gray whale*, the Latin term *Eschrichtius robustus*, and the "Makah" term, *Lunch*.

I do not think these moments represent a special desire on the part of Makahs to tell people about their dietary preferences *tout court*. Food, perhaps more than any other category, straddles the boundaries between salient aspects of life on the Makah Reservation—connecting spirituality and economics, sociality and decorum, ethics and politics. So it is not surprising that people in the Makah community express their opinions on whaling in the idiom of food. But it is also an open secret on the reservation that gray whale meat is not particularly tasty. Makah whalers consider it inferior to humpback whale meat in every way. "Why have hamburger when you can have steak?" is something whalers frequently ask when comparing the two. The reception of gray whale meat into the community after the 1999 hunt was mixed. I heard some stories about whale teriyaki and smoked or canned whale meat, but few people clamor for its reintroduction into their diet today. Soon after the tribe announced its intent to revive whaling, federal agents allowed a Makah fisherman to keep and butcher a gray whale that had become entangled and drowned in his marine set-net and to distribute

Figure 2.1 The Makah word for "gray whale"
Photo by the author; unknown artist

the meat to members of the community. It was the first time a tribal member had been allowed to do so in over eighty years. Accounts of the butchering suggest a kind of curious culinary adventurism, a performative meal motivated by political aspirations rather than the other way about.[1] There are surely Makahs who relish the flavor of gray whale meat and wish for it to be available to them; Theron Parker took the opportunity to butcher and distribute a gray whale that washed up on the shore of the Olympic National Park near the Yellow Banks in 2001, for instance, and claimed the meat and blubber fed more than 100 families. My point, however, is that a longing for the taste of gray whale meat is not a primary motivating factor in the tribe's return to whaling.

Why, then, do people go out of their way to foreground the edibility of gray whales in conversations with strangers or with one another on social media? And why are such assertions so often inflected with an emphatic and even hostile tone?

To designate a living being as *food* is a moral distinction, something that the analytic philosopher Cora Diamond points out with deceptive simplicity in her 2004 essay "Eating Meat and Eating People." Ethical vegetarians who admit they would eat an animal that had died an accidental or painless death—struck by lightning, for instance—show us how durable such categories can be. For Diamond, the things philosophers say when they attempt to justify animal rights matter less than the categorical distinctions people make between, say, livestock and pets, or animals that can be eaten and animals that cannot. This is something I noticed again and again in interviews with members of the Makah community whenever I asked whether some attribute of whales (like their putative intelligence) should factor into discussions about hunting them. "We eat them," Makahs would say to me with a baffled expression. "We've always eaten them."

To consider gray whales as food is not only a moral distinction, it is a moral distinction that runs against the grain of majority opinion in the United States. This is especially true in the Pacific Northwest, the birthplace of antiwhaling activism in North America, where whale watching is a multimillion-dollar industry and where affectionate images of whales adorn public spaces in every urban center. The Pacific Northwest is also a geopolitical region with a fraught history of settler colonialism relating to indigenous hunting and fishing. In the early twentieth century, Makah fishermen were often vilified by the state of Washington as poachers and threats to shared natural resources. The aftermath of the Boldt decision stoked resentment that still simmers today. Some non-Native fishermen view the ruling as having established a racial double standard in favor of Native fishermen, who are allowed to fish outside of regular seasons and take fish in excess of the quotas set by conservation biologists. And when indigenous fishermen are not vilified, they are sometimes treated as ecological saints and held to an unreasonable and fictitious standard that accords more closely with Western environmentalist tropes than with any historically accurate indigenous way of life. The "Yum!" response plays upon the moral connotation of eating whales to push back against both of these misrecognitions by defiantly rejecting mainstream views and deliberately irritating racial tensions. It is as if to say, "Yes, we kill animals, do you have a problem with that?"

For all its political value, however, the "Yum!" response (hereafter, *yum*) has as its basis a real moral difference between whalers and antiwhaling activists. The edibility of whales is one of the few points on which I am comfortable arguing that there is such a thing as a (mostly) unified and distinct Makah point of view. *Yum* may be politically motivated most of the time and may be exaggerated to the point of self-parody, but the sentiment is not insincere. A cultural logic underlies the response that is concerned with more than provocation.

Eating Well in Neah Bay

Even if the exclamation of "Yum!" or "We eat them!" is never really—or rarely entirely—about a desire for gray whale meat, the statement draws much of its meaning from what it means to eat well in Neah Bay. That culinary language would be the preferred register for this kind of assertion feels overdetermined, and there is something to be gained from thinking about why.

Food—talk about it; access to it; and the preparation, distribution, and sharing of it—are pervasive features of daily life on the Makah Reservation. Food shares a porous boundary with nearly every significant aspect of life in the community, from the partitioning of time in public space to religious practice to the maintenance of family relationships. Every community event I attended began with the serving of free food provided by the host family or the tribe. Often, the period reserved for eating greatly exceeded the length of the rest of the event.

Two demographic characteristics that are emphasized in almost every description of life on the Makah Reservation—the level of poverty in the community and the exceedingly high rate of unemployment, which by some estimates is as high as 70 percent—are deeply problematic because both measures miss the importance and prevalence of food items in forms that never enter circulation as commodities. In her doctoral dissertation, political ecologist Jennifer Sepez (who later worked for NOAA) found that subsistence items accounted for as much as 70 percent of the meat consumed by tribal members in 1998 (Sepez 2001, 130–33).[2] As Sepez notes, these subsistence foods link modern Makah households through resource-sharing arrangements that in some ways mimic precontact forms of sociality (267–76).

Commitment and skill in gathering subsistence foods like mussels and clams is widely seen as a virtue in the community and a means of distinction. A local saying goes, "When the tide is out, the table is set." One of the things I missed most after leaving Neah Bay upon the completion of my fieldwork was this shared awareness of the rhythms of the tide—the excited talk at the local restaurant, for instance, of the next big minus tide that would expose the mussels and oysters and clam beds for gathering.[3]

Ethnographic literature on Northwest Coast societies is replete with examples of the strong connection between food and religion.[4] As Greg Colfax, a Makah elder and self-taught scholar of comparative religion, explained:

> There are so many rituals and so many events that individuals have acquired in their quest for spiritual power, so much of it is connected to getting food. "With this power you'll be able to be a good hunter, or you'll be able to know

where these seals live, or you'll just know how to get more food" . . . I guess what surprises me is that the Christian religion is not really about food. It's . . . it's more social. And the Muslims, it's about justice. Buddhists, it's about ending pain. . . . But with us, it's about food. . . . The rituals have so much to do with food. . . . No matter which [ritual] you look at, tacked onto some place is this: "Because I do this, I have the ability to get more food."

The enormous quantities of blubber and meat provided by a single whale enabled whalers to maintain their high social status, but an animal that provided so much food in a single stroke also crossed over into the realm of the sacred, adding to the spiritual significance of its pursuit. After reflecting on the deep spiritual aspect of the 2007 rogue hunt, Theron Parker went on to describe the hunt to me as "basically just grocery shopping." A prototypically mundane activity in many other American communities, grocery shopping thus described illustrates the lack of a distinction between things-that-are-sacred and things-to-be-eaten.

For all the emphasis on food gathering in the community, however, it *is* possible to overstate the importance of so-called traditional foods. Even those people who are celebrated for their skill in gathering such foods tend to rely heavily on supermarket fare, which is commonly referred to metonymically as "Costco." Generally speaking, people readily take advantage of the cheap retail meats available in nearby cities. Potlatches and other community events are more likely to feature beef, potatoes, green beans, and other standard working-class American fare than local seafood, despite the abundance of fish at the nearby docks. And despite the hypercognized status of food for locals, visitors are likely to experience Neah Bay as something of a food desert. With the exception of local smoked fish, processed foods dominate the local culinary landscape. Fry bread and Indian tacos (two pieces of fried dough filled with seasoned ground beef) are widely consumed and commonly advertised as "traditional foods" at roadside eateries. Some members of the community blame these unhealthy foods for the prevalence of diseases like diabetes. Access to healthy food sources was one of several publicly stated justifications for the return to whaling. Generally speaking, though, these different kinds of traditional foods—locally harvested fish and shellfish and seal oil, on the one hand, and fry bread derivatives, on the other—are not commonly distinguished in practice.

Yum is a kind of total social fact that refracts all of these considerations, pointing toward the significance of food and eating, the traditional status of whalers, the tribe's history of colonial oppression and struggle, and the opportunities and challenges of the contemporary food landscape.

The Politics of *Yum*

Uttering "Yum!" in response to a question about gray whales is politically effective precisely because it refers not only to cuisine but also to edibility as a moral classification. As a blunt expression of Makah views about animals, it is meant to shock and offend a perceived animal-loving public, marking a refusal on the part of the speaker to affirm the (dis-)tastes of the white majority. This is why people can say "Yum!" even while acknowledging that gray whale is not particularly tasty, or "We eat them" even when that has not been true for fifteen years (or for the better part of the previous century). The fact that *yum* is something of a silly English word only makes it a more effective vehicle for this kind of self-parody as a criticism of the non-Native American mainstream.[5]

Edibility itself is a metonym here in its turn. *Yum* stands in for a distinctly Makah attitude toward animals, an attitude I would describe—with the exception of pet ownership (more on that later in this chapter)—as utilitarian, pragmatic, and decidedly, often stylistically unsentimental. One of the first people I talked to during my initial visit to Neah Bay was Donnie Swan, a tough young whaler who served as the second harpooner on the 1999 crew. In our first interview, he spoke matter-of-factly of killing seventeen bears in the river near his home over the past two years because they had been coming too close to his fishing nets. "I shoot every [bear] I see," he told me. "I'll give 'em away or make sausage out of them or something." Five years earlier, the tribe had issued a bounty on bears because, as he told me, they "were rippin' up so many trees and stuff, they was givin' us fifty dollars for a pair of ears."

Donnie was extreme in his expressed disregard for bear life; many Makah fishermen simply avoid the areas where bears are known to be. But Donnie's outlook did not differ dramatically from the attitudes of many people in the community, especially fishermen, who view animals through a utilitarian lens that tends to agree more closely with the working-class ethos of nearby logging towns than with environmentalist stereotypes of the "ecological Indian."

Jerry Lucas, the charismatic erstwhile director of Makah Forestry Enterprises, likes to tell a story in which he hints at the intended effects of sharing such utilitarian attitudes toward animal life. In the 1980s, Washington State fisheries officials were plagued by Herschel, a gluttonous California sea lion who installed himself near the fish ladder at the Ballard Locks in Seattle during the annual steelhead spawning migration. Herschel singlehandedly "devastated" two steelhead runs, Lucas told me, but since sea lions were a federally protected species, state officials were reluctant to kill him. Instead, they attempted to relocate Herschel to the Pacific Coast not once but twice. Each time, the resourceful sea lion found his way back.

"They *flew* the thing out," Lucas told me and a group of Makah whalers over coffee one morning, sounding both angry and amused. "It swam back in five days! And the second time, it only took three!"

Later, Lucas told me how he would have handled Herschel:

> I was with the tribal attorneys when they first broadcasted that [story about Herschel]. I was working on a timber issue. And the young attorney then was, uh, named John Arum. "Lucas," he says, "we need to take care of this." I said, "John, get a hold of the Washington State Department of Fisheries, you tell them I will take care of that problem *this evening*." He goes, "Really?" Course, he didn't quite—wasn't up to snuff on Makah yet, cause he's learnin'. So he looked at Al—the main man who started that firm was named Al Ziontz. So he's sitting there going, "Lucas, don't go there!" I said, "Al, I'll go shoot that sumbitch tonight. And tell the state, we'll only charge them a hundred thousand to kill it today. But after today," I said, "it's going to cost them a million!"

The neophyte attorney stands in here for an easy-to-offend animal-loving public. The story's running theme is that Lucas himself would never commit the resources to evacuate a sea lion, and his preferred solution, which he implies is distinctly Makah, would be irreverent, lethal, and out of step with the general sentiments of non-Indians in the region.

Lucas's story is one of many instances I saw where Makahs pushed back against the romanticism of indigenous beliefs. "You got all these people," Janine Bowechop, the director of the Makah Cultural and Research Center (MCRC), once said to me, "we kind of burst their romantic bubble. You get these people who said, 'I thought you people worshipped animals and the wind and prayed to the four directions' . . . you kind of can figure out an image that they have of what Indians are supposed to be like and, um, it just almost makes you laugh."

Makah officials tend to be less bullish in their articulation of this countermainstream sentiment. They see less value in the repeated and emphatic articulation of the ethics of *yum* and tend to be more concerned about how Makah individuals' matter-of-fact statements regarding animals might be perceived by certain publics. When the MCRC Board, a body that also serves as the de facto local review board for academic research, met to review my research proposal, one board member expressed concern:

> Part of what bothers me is what you're talking about with your "moral dimension" is *Bambi*. You're talking about [a] society that's been affected by *Bambi*. And a lot of *our* society doesn't think like that. Our children certainly watch *Bambi*, but they're also taught our history and our culture. So we don't look at *Bambi* the way a non-Indian would.

I had described my interest in the moral dimension of whaling much as I would have in an academic setting. I meant for the word to be understood in the descriptive sense; I was not interested in passing judgment myself, nor is that the aim of this project. My choice of words raised concerns for the board member, however, who feared that my project would be tantamount to a moral judgment. She suspected that the criterion for that judgment would be the particular moral landscape of the contemporary United States as she imagined it to be: an empathic, anthropomorphic, nonutilitarian ethos that she abbreviates here eponymously as "Bambi." The board member, the head of an influential Makah family whom I came to deeply respect in the time of our short acquaintance, later told me a story about how seeing *Bambi* had "ruined" her brother for hunting when he was very young; he would never go hunting with their father again after seeing the movie.

"Whereas you see Bambi—," I began to ask, and she finished my sentence with gleeful flourish: "Meat!"

The Ethics of *Yum*

It would be tempting to focus entirely on the political mobilization of *yum* because it is interesting and because morality—despite a resurgent interest within the field of cultural anthropology—is still arguably a less fashionable topic today than power or politics per se. But the political deployment of *yum* starts from a set of assumptions about the ethics of human-animal interaction widely shared and reiterated in the Makah community, and this is worthy of some elaboration. These assumptions take their plainest form in an unlikely place: the formal statements of whaling opponents within the Makah Tribe.

It was June of 1996, just over one year after the Makah Tribe publicly announced its intent to hunt gray whales after a seventy-year hiatus, when seven Makah elders took out a full-page advertisement in the *Peninsula Daily News* condemning the proposed hunt (see figure 2.2).[6] Among them was a quartet of highly respected women in their late eighties and early nineties

> We are elders of the Makah Indian Nation (Ko-ditch-ee-ot) which means People of the Cape. We oppose this Whale hunt our Tribe is going to do.
>
> The opposition is directly against our leaders, the Makah Tribal Council, Tribal Staff, and Bureau of Indian Affairs, which is an arm of the United States Government.
>
> The Makah Tribal Council has been functioning without a quorum; two Councilmen are off on sick leave for very serious reasons, cancer.
>
> How can any decision be legal when our by-laws state the Treasurer shall be present at every meeting? The Vice Chairman is the other man out.
>
> The Whale hunt and other important issues were never brought to the people for a vote, or simple notification.
>
> The Whale hunt issue has never been brought to the people to inform them and there is no spiritual training going on. We believe they, the Council, will just shoot the Whale, and we think the word "subsistence" is the wrong thing to say when our people haven't used or had Whale meat/blubber since the early 1900's.
>
> For these reasons we believe the hunt is only for the money. They can't say "Traditional, Spiritual and for Subsistence" in the same breath when no training is going on, just talk.
>
> Whale watching is an alternative we support.

Figure 2.2 Statement by "elders of the Makah Indian Nation," as printed in the *Peninsula Daily News*, 1996

known locally as the Old Ladies. Accomplished in traditional arts, they were among the last native speakers of the Makah language. "We are elders of the Makah Nation," the advertisement began. It went on to expressly question the purpose and legitimacy of the hunt. Citing a lack of proper spiritual preparation and training on the part of the would-be whaling crew, the signatories concluded that the hunt was being pushed through by tribal leaders without a quorum and would be conducted "for the money."

Fifteen years later, I spoke with a woman I will call Esther, a daughter of one of the signatories to the letter. Esther's mother had traveled to the 1996 IWC meetings as part of the antiwhaling delegation. "We didn't eat it anymore," Esther told me as she explained her own reasons for publicly opposing the hunt. "My mother said that she didn't eat [whale] as a young child, and so she said we've gotten along without it all this time, she didn't think we needed it. She mainly didn't think that we would eat it. Who would know how to cook it?" She concluded the thought by stressing that "if they were gonna eat it, that's fine, that's different. It'd be okay if they were gonna eat it. But to just go for the sake of it . . . just so they could say 'Look what we did.' That's not okay."

As it happens, "Look what we did" is actually a fair if somewhat ungenerous description of how a number of Makah whalers justified the first whale hunt. Whaling was, among other things, a political performance meant to show the world that Makahs could not be bullied or stopped from reclaiming sovereignty over local marine resources. How willingly local people accepted such a political rationale was attributable in part to generational differences and the extent of their own personal involvement in the fight for treaty fishing rights from the 1940s to 1960s.

Regardless, what I want to point out here is that the objections voiced in the elders' advertisement and in Esther's comments take aim at what constitutes a proper *use* of whales. The edibility or usability of whale bodies is never questioned. If they were going to eat the whales, Esther emphasized, then it would be perfectly okay to go whaling. The elders' advertisement thus affirms the dominant view of whales in the Makah community as effectively as any Makah whalers' manifesto, mostly by what it does not say. Although several of the elders who signed the advertisement would later travel to International Whaling Commission meetings in Scotland and Monaco under the banner of Sea Shepherd Conservation Society and other antiwhaling groups, the advertisement sounds none of the familiar preservationist themes advanced by those groups. Instead, the elders' stated objections are largely procedural, and their statement gives no indication that any of the elders believed killing whales was wrong per se. Their objection was precisely that whales were meant for eating, not for political games.

Binki's Challenge

By all accounts, Alberta "Binki" Thompson initially opposed the revival of whaling not out of any special love for whales but because she felt the tribe no longer needed the meat. "I don't need to kill a whale," she was reported to have often said, "to know that I'm Makah." An elder and a grandmother who was known to some as a bit of a gadfly, Thompson worked part-time at the tribal senior center prior to the first whale hunt. According to witnesses, her reasons for opposing whaling changed when she visited the birthing lagoons in Baja California and was approached by a mother gray whale and calf. Following the encounter, Thompson said she felt she had developed a special connection with the whales that animated her opposition to whaling. "My dream," she said, "is that I wake up one morning and the tribal council has called a conference to make a statement: We now realize that this whale gave up its life for us a hundred years ago so that we could eat. Now we want to honor and protect the whale until the end of time."[7]

As the highest-profile of the Makah opponents of whaling, Thompson traveled to IWC meetings around the world and became closely associated with antiwhaling activists like Paul Watson and Margaret Owens. By the time I arrived in Neah Bay in 2010, she was living in an extended care facility in Port Angeles, a two-hour drive from the reservation. Her beliefs, motivations, allegiances, and genealogy all became highly contested in the years since the height of her activism. Having never met her, I can only comment indirectly, as if she were a character in a story with many retellings. That story suggests that Thompson's opposition, although not without worldly benefits, also came at a considerable cost to her. Thompson lost her part-time job with the tribe after speaking out against whaling, and tribal leaders threatened to banish her from the community if she continued to speak out.[8] According to Thompson, the Makah chief of police delivered a written ordinance specifically prohibiting her from speaking out against whaling (Hogan 1995). Shortly before the 1999 hunt, Thompson's dog was taken from her home and killed under mysterious circumstances.[9]

Whaling supporters within the tribe sought to undermine Thompson's authority first by physically removing her from the community and then, when that failed, by drawing upon the memory of older social distinctions. In Neah Bay, the accusation that one's political opponents descend from "slaves," and thus have no legitimate authority in the community, seems never far from the lips. Robert Sullivan describes this "highest insult" that was in the atmosphere during the hurry-up-and-wait period before the fall 1998 hunts:

> While the whale hunt was on hold, while the political alliances that had made a whale hunt possible strained in the wait, that word—*slave*—was being whispered around Neah Bay. People disputed birthright and lineage and family trees, the right to hunt a whale, the right to speak for a family, for a village, for a tribe. *Slave*: it was being whispered like a dirty word. Slave and chiefs, whalers and commoners: everyone seemed to have a different idea of who was which. (138)

A different journalist paraphrased Dan Greene's thoughts on the matter as follows: "There were nobles . . . and there were commoners. Thompson . . . belongs to neither class. That leaves only slaves. Given her social status, Greene believes Thompson's opinions should properly have been ignored" (Blore 1996).

Thompson's story is exceptional and highly visible, but it is by no means the only challenge to the orthodox Makah view on whales and whaling. Lesser challenges are managed without resorting to spectacular threats of violence and banishment. There is a kind of self-policing that takes place as a result of the political implications and sensitivity surrounding talk about animal ethics in the community.

Two subtle examples should suffice to show how people police their own impulses with regard to the politics of *yum*. In the first, Jean Vitalis, the museum board member who chafed against the moral economy of Bambi, placed herself in the small minority when she criticized (to me, in an interview) the killing of bears and cougars on the Makah Reservation by tribal members:

> Um . . . I think it may . . . I . . . I . . . I think it just makes me feel bad that they're . . . I mean, they didn't . . . I don't know. They, they have a right to be where they're at, I guess. I don't see why you should kill 'em if you're not gonna do anything with 'em.

Normally direct and very well spoken, Jean struggled to articulate why she might oppose the killing of these predators. On two different occasions, Keith Johnson brought up the practice of clubbing baby seals in the Canadian Arctic as something he was "sensitive about," something he might "vote against" if given the opportunity. "But then again," he said, reining in his judgment each time, "I'd have to look at the numbers of [seals] and how they are impacting fish."

Keith and Jean both stop short of condemning practices that clearly bother them, and both manage to recast their complaints in locally approved ways of thinking about animals. Jean ultimately suggests that killing bears or cougars is wasteful, while Keith brings the conversation back to population ecology. Their reticence shows that it is difficult—I am tempted to say impossible—to express the sentiments that they attempted to convey in a way that is consistent with the views of whales and whaling I have described in this chapter. Prevailing attitudes in the community allow for the killing of animals without social sanctions, provided the act is not wasteful or contrary to some other principle of respectful wildlife management—a set of attitudes neatly expressed by the commonly repeated phrase, "Take only what you need."[10] To tell a member of the community that killing bears or cougars is intrinsically wrong, in this context, is to say something nonsensical. Such thoughts are relegated to the realm of the idiosyncratic, even the heretical, and expressed with commensurate diffidence.

None of this is to say Makah whalers do not feel ambivalence or remorse after a successful hunt. As Theron Parker told me:

It's just another creature, but it's a big creature, and it's like, you know it's in pain, you're trying to—what do you want to say, euthanize it—as quick as you can. Y'know, the benefit we're gonna get out of it is more than the heartache we're gonna do to take it, but if you don't have any remorse then what kind of person are you? For anything, really, if you're gonna take a life.

A Whale on the Beach

Social orders cannot maintain themselves by the threat of force alone, and self-censorship has its limits, too. Makah whaling would not be viable (as a political gambit or anything else) without the local popular sense that whales are things-to-be-eaten. I was living on the Makah Reservation for a surprisingly short time before I, too, began to see gray whales differently.

On a frigid February morning a little more than one month into my fieldwork, Wayne Johnson handed me a compact disc with images from his trip to the Russian Arctic, where he and another Makah whaler had hunted gray whales with Chukotka Natives on the Chukchi Sea. Wayne and I had become acquainted during my short visits to Neah Bay the previous year, and I would eventually spend many hours with him during my fieldwork, digging clams, drinking coffee, fetching skiffs from this place or that on the river, or just riding around town in his rickety old pickup truck. Wayne introduced me to many of my eventual informants, and he made me feel welcome at morning coffee sessions with a group of his associates and fellow members of the whaling commission. Robert Sullivan (2000) described Wayne as "five foot eleven and wiry, with a chin that he holds up high in unconvincing defiance whenever he gets a little nervous" (43). I found the description accurate, if incomplete; Sullivan never fully related the vulnerability and—for lack of a better term—sweetness that make Wayne a sympathetic figure despite the rough edges around his character.

As we flipped through some of the images, Wayne told me about the Russian hunt and about how the Makah Tribe had paid for him and Donnie Swan to travel there. He told me how extreme the poverty was in the Chukotka community, how desolate the landscape was, and how the kids there chewed baleen like it was candy. He confessed his relief that after a few days they had given him something to eat besides whale meat.

The photos of the whale hunt that he shared with me were bloody and graphic. The photo roll began with smiling faces at an airport in Alaska, then aerial shots over a frozen landscape, followed by shots of small powerboats buzzing through foaming water, and, finally, whales. Whales that had been dragged up by heavy equipment and laid upon the beach in rows, stomach contents hanging out, large chunks of flesh missing from the head. Whales half butchered, their eyes half lit, little rivers of red blood running through the mud next to their massive bodies.

There is much to be said about the Makah whalers' trip to Russia, but on that February morning I was taken aback most by my own reaction, or nonreaction, to the images. From the outset of my fieldwork, I had been resolved to pay close attention to my own evolving sentiments toward whales and whaling during my time living in the community—to use my own shifting perceptions and visceral reactions as an ethnographic tool. As I became acquainted with whalers as well as antiwhaling activists, I wanted to examine how my exposure to their polarizing views would impact my own attitudes and feelings. Viewing the photos was the first time I noticed such a shift. I noticed that despite the carnage, despite the many massive creatures lying on the beach next to piles of their own entrails, I felt nothing. Just a few years earlier, when Animal Planet's *Whale Wars* first appeared on television, I refused to watch the show, concerned that I would become too upset. I didn't know the difference between minke whales, pursued by Japanese whalers in the TV show, and any other species. Like most Americans, I was uninformed about the state of global whaling. Despite my lack of knowledge, I nonetheless opposed whaling. Whales, to me, were giant, gentle, intelligent creatures that we ought to communicate with rather than kill. To drive harpoons into their bodies just to make oil out of them seemed gratuitous and wrong. My feelings had tempered significantly in the time intervening, such that I felt comfortable undertaking this project without judging either side in advance. Still, I was surprised by my nonreaction to the images. Just a few days earlier, I had felt (or not felt) something similar when I noticed a grainy image of a half-butchered whale hanging in the trophy case at the Neah Bay High School cafeteria. The whale looked to me like a hunk of meat. More to the point, the whale looked to me to be in its *natural* form. If a few months of living in Neah Bay, spending time with whalers, and hearing their stories could have this effect, then it suddenly became very clear to me how, if I had been raised in Neah Bay, heard songs of whales offering themselves to feed the village, and told stories about ancestors who made whales "come home" by offering themselves to hunters, I would see the natural image of a gray whale as *on the beach.*

For most Americans, the prototypical form of a whale is probably the spectacle of a breaching humpback, immortalized in advertisements for Pacific Life Insurance. For Makah whalers, the prototypical form of a whale is heaped upon the sand above the line of breaking surf. A multitude of factors

Figure 2.3 A whale on the beach, ca. 1910
Photo by Asahel Curtis

conspire to make this so, subtle mechanisms that work upon the mind to create an image of whales as *on the beach* even before they approach a whaling canoe. The phrase itself, "a whale on the beach," shows up frequently in everyday conversations. Whalers are fond of using the idiom to lament delays in the administrative reapproval process. "We should have twenty whales on the beach by now," someone might say. "What would the government say if they came down here and we already had a whale on the beach?" Wayne Johnson never tires of asking. Most surviving photographs of early historic whaling show whales either on the beach or floating in shallow water. A common theme in Makah spiritual discourse has whales "coming to the beach" or "coming home." Old whaling songs and prayers speak of bringing whales "to the beach." The tribal chairman at the time of my research spoke often of his renowned ancestor, a whaler whose name roughly translates to "he makes the whale blow on the beach."

For many of the tribal members with whom I spoke, the whale being dragged onto the shore of the Makah Harbor during the 1999 hunt was the most memorable moment of the last two decades. They spoke of the event in rapturous tones and described an overpowering collective effervescence.

Figure 2.5 Thunderbird and the whale

Even the tribe's artistic motifs work to build this prototypical image of the whale. The tribe's emblem features a bewildered-looking whale in the clutches of the primordial god Thunderbird.[11] The image offers a symbolic equivalent to the phrase "a whale on the beach." The iteration of this idea in all its forms produces the prototypical, desired form of the whale as already dead, already *food*.

The (Other) Politics of *Yum*

If this story were to end here, it would largely have been one of empowerment, of tribal members leveraging moral difference to positive effect in a field of opinion where cultural difference is incentivized. Even the attempts to silence dissenters like Alberta Thompson were made to satisfy the expectations of the International Whaling Commission, who, however misguided in their understanding of culture, were willing to issue an exemption to the moratorium on commercial whaling on the basis of "cultural need."

Of course, there is a darker history to the attribution of certain kinds of environmental ethics to Makahs and other Indian groups, one in which cultural difference did not carry a positive valence. Throughout the early twentieth century, representatives from the state of Washington portrayed Makah fishermen as poachers threatening the health of fisheries, especially the steelhead runs cherished by white sport fishermen. This vilification led to the exclusion of Makah fishermen from their usual and accustomed fishing sites on the Hoko River and elsewhere, which in turn led to the civil disobedience I described in chapter 1. The revival of Makah whaling stirred up similar racialized resentments over marine resource rights, some of them stemming from the outcome of those controversies over the management of salmon and halibut. As a result of all this, statements about Makah attitudes toward animals are rarely politically neutral but are instead freighted with this overlapping history of colonialism and romanticism. Sometimes, when I asked a member of the Makah community about their attitudes toward whales, it was suddenly as if we were no longer alone, and the state of Washington and all of its non-Native residents were audience to the reply.

The pejorative stereotype of the "poaching Indian" haunted one exchange I had with Jo Jo McGimpsey, janitor of Neah Bay High School and a recognized expert on what someone who recommended him called "the spiritual side of things." Gregarious and warm, Jo Jo has a scornful side that he reserves for people he describes as "fucking Republicans" and anyone he suspects of being a "whale hugger." Jo Jo regularly emcees community events like potlatches and naming ceremonies, and he is commonly called upon to sing or deliver prayers at public events. Jo Jo served as a sort of consultant to the 1999 tribal whaling crew, helping them prepare spiritually for the hunt. When the whale in the 2007 rogue hunt slowly died of its injuries (the whaling crew had been arrested by the Coast Guard hours earlier), Jo Jo was taken out by boat to pray over the dying whale. He reportedly sang over the whale for several hours.

Jo Jo was one of several Makah men who chided me for using the word *kill* in reference to Makah hunters taking animal lives. For that reason, our first formal interview began with him seemingly on the defensive. "We hunt to provide for our people, our families," he began. "White people, they just kill and poach because they think they're above the law. We hunt. We provide. We don't just go out and kill things. There's more white poachers than they know of, that's been convicted, too. We hunt. We provide meals. Big difference, *kill* versus *hunt*."

When I tried to clarify what he meant by "poaching," he quickly responded:

[Poaching is] the way you're phrasing it. Just killing stuff. You kill things. We don't just kill things. We hunt. We provide. Big difference. *Big difference.* And if you're gonna say I say I enjoy killing things, no no no no no. I'll object to that and I'd sue you. Cos I don't kill things to kill things. There's a reason why you take a life. . . . We don't just go around killing things. That's what the white guys say [about us].

Jo Jo is shadowboxing with a pejorative stereotype that haunts white-Indian relations in the US Pacific Northwest, one born of the same tensions that fueled the civil disobedience of net bans in the mid-twentieth century and Native "poaching" before that: some people in the region harbor the perception (or are perceived by Natives to harbor the perception) that Native Americans have a penchant for killing animals illegally and indiscriminately.

Throughout our interview, and perhaps with this perception in mind, Jo Jo insisted that there always needs to be a "good reason" to kill. Given that insistence, I was interested to hear about his experience as a commercial fisherman, particularly how he navigated the common problem of seals taking fish off of hooks before the fishermen could retrieve them. Seals are loathed by most Makah fishermen in part because they have a habit of taking a bite out of the protein-rich, egg-filled bellies of the salmon, thus rendering the entire fish commercially worthless. This had always seemed to me like a conundrum: the rejection of the word *kill* on the one hand, and the almost universal loathing of seals for commercial reasons, on the other, which often led fishermen to kill the seals and discreetly dump their bodies on the shore.

The exchange that followed my question underscores the limitations of the romanticized view of Native American attitudes toward nonhuman animals, but it also demonstrates how the specter of the Poaching Indian endures in the minds of people in the community. To preserve its nuance, I have included the entire segment of the conversation here, uninterrupted:

Les Beldo: When you were a fisherman, did you shoot seals?
Jo Jo: (*incredulously*) Did I shoot seals?
LB: Yeah, I know some fishermen will shoot seals when they get too close to their line.
JJ: Well, yeah, I'd shoot 'em if they was takin' fish off my line. Yeah. Hell yeah. Worthless, no good. Just like a cat, all they do is eat, swim, and shit. Cats, all they do is eat, sleep, and shit too. Seal's the same thing. Worthless. Ah, I shouldn't say worthless. Harbor seals are good for seal oil. But the Steller and

the California [Sea Lion], yeah, hell all they do is eat all the salmon, fish. Not good for anything.

LB: So you wouldn't feel bad about shooting them and not eating them?

JJ: Hell no. Cause they're a nuisance.

LB: Are there any other animals that are like that?

JJ: That I just kill to kill? I don't kill to kill.

LB: I didn't say kill to kill. [Animals] that are a nuisance.

JJ: That are a nuisance? Skunks. I got one under my house, I think. . . . Yeah, a nuisance. Cats are a nuisance.

LB: So if you could catch that skunk, would you kill it?

JJ: Hell yeah! Get it the hell out of here, I don't want it coming back. He'd come back pissed at me and squirt me more. Hell yeah, I'd kill it. Rats. They're . . . God, they're ugly. Rodents.

LB: What about bears? There was a bounty on bears here a few years back.

JJ: Hm hmm. [Yes]. If you're—there, again, you have a reason. The reason why they put a bounty on them, y'know, and it's an accepted procedure. So some folks go out and hunt 'em, just for the bounty. But you have a reason. You don't just go out there in the woods, "Oh, man, there's one," bam, and leave it. That's not good. Always, there's always a reason for something.

LB: Even if you're [just] protecting your commercial fishing line from a seal, you think that's a good reason?

JJ: Good enough for me. Damn right.

LB: Would you say then that seals aren't as respected by fishermen? 'Cause you talk about having respect for all animals, but a lot of fishermen have the same kind of attitude, "Just get the damn things away from our gear."

JJ: Sure. Yeah, they're . . . useless. All they do is eat fish.

LB: By the same token, I guess, you could say that about orcas, or about killer whales, 'cause they eat tons of salmon.

JJ: Yeah, I don't see nothin' wrong with killin' 'em.

LB: With killing orcas?

JJ: Yeah.

LB: Really?

JJ: They're just a big fish. What makes 'em so special? We used to [target] practice on 'em! . . . I mean, I would only imagine, would only imagine, there's probably some good medicine in 'em too, ooh. They're also involved in our legends and tales, so there's always a connection.[12]

LB: But there's people who would harpoon them just for practice and then not even . . . would they keep them?

JJ: Oh, I'd imagine so. Get something out of them. Teeth [maybe].

LB: So what, then, what is not a good reason to kill? If you have no reason, and you just do it for the fun of it, you're saying that's not a good reason?
JJ: Nah, that's indiscriminate! You don't need to take a life just to do it. Crazy people do that.

Then, at the end of the interview, Jo Jo made a final extended comment on my use of the word *kill*:

> When you first called me, all I heard was the word *killing*. Killing, killing, killing animals. To me, you sounded like a caveman . . . pretty archaic. If you're working towards your doctor [sic] the people will look at you and say, that Les, he's pretty rough-spoken there. And that might turn some people off. [It] almost turned me off, but I figured, I says I'm not here to judge this young man that wants to be a doctor and is working hard, if this can help him, that's what he needs. So, we don't kill just to kill, and everything has a reason and, I mean it goes back, everything is connected. Everything has a spirit or it's alive or there's something that's—a rock is not just a rock.

At first, it is difficult to make sense of Jo Jo's comments. Despite his admonishments over my use of the word *kill*, he speaks of seals and other "nuisance" animals in a way that does not seem reverent by any definition of the word. His preferred expression, *ask the animal for its life* instead of *kill*, seems inappropriate for those instances where a seal or skunk is shot for being a nuisance. Most of all, the point of having "good reasons" to kill is unclear when almost anything seems to qualify as a good reason, including the opinion that rats are ugly and that skunks smell bad.

Jo Jo's seemingly contradictory statements might exemplify how lived realities matter less than the ideal when the discussion turns to ambivalence over killing animals. Wendy Doniger (2014) has speculated that all human beings feel ambivalent about killing animals and that prohibitions against certain forms of killing help to overcome that ambivalence when one does take a life. The idea that animals sacrifice themselves and are happy to die, Doniger argues, allows humans to figuratively "have one's animal and eat it" (414). For instance, on two separate occasions I told Greig Arnold that I always felt conflicted whenever I killed a white-tailed deer (the largest animal I hunted as a teenager), and each of these times he gently admonished me for feeling that way. "You ask for that deer's life next time you go," he told me. "You see what difference that makes in your—if nothing else, understand [the deer] has a family, and do a little ceremony thanking it. Oh, you watch, the difference in how you feel. Your inner self will be more . . . settled. That's what I think, anyway."

Jo Jo's comments are impossible to understand, however, without taking into account the interactional effects between a white ethnographer who asks conspicuous questions about killing and a local expert—but not a recognized tribal spokesperson—who is deeply aware of the criticisms of Indians in the region.[13] Jo Jo is responding to two persistent misrecognitions: the pejorative stereotype of the poaching Indian and the romanticized ecologically noble Indian, the latter of which is now much too close to the animal rights ("whale hugger") view exhibited by antiwhaling activists to be fully embraced by Makah whalers.[14] I had already primed this defensive posture by overusing the word *kill* in my previous correspondence with Jo Jo, and my question about shooting seals set the tone for what followed. No other interview I conducted unfolded like this, although Greig Arnold also chastised me for using the word *kill*. "We don't kill," Greig once told me as we sat in his office at the tribe's fisheries management department. "We ask the animal for its life."

Whales as Persons

The attitudes toward whales and other animals that I have described in this chapter are not hidden from the broader public; on the contrary, sometimes they are flaunted for rhetorical effect. Despite this, and despite all that historians have documented on the variegated history of indigenous ecological thinking, myths about Native American attitudes toward animals persist.[15] It is said, for instance, that all Native Americans revere animals (Hughes 1983); that they use every bit of an animal out of an inherent sense of conservationism (Smithsonian National Zoo 2014); that Native peoples have a special relationship with the land and live in balance and harmony (Sponsel 2012); that "the hunter-gatherer mind is humanity's most sophisticated combination of detailed knowledge and intuition" and "the hunter-gatherer seeks a relationship with all parts of the world that will be in both personal and material balance" (Brody 2000, 292). When indigenous peoples conduct themselves in a way that does not align with these stereotypes, accusations of corruption and inauthenticity often follow. "Even well-meaning people [create] images in their minds of what Indians represent, and it usually involves worshipping all mother earth's creatures, praying to the winds and maybe living off of vegetables and roots," Janine Bowechop writes.[16] Paul Nadasdy (2005) attributes this problem to what we might call the failure of the Western environmental imagination, noting that within the normal spectrum of Euro-American environmentalist possibilities, people tend to treat reverence for nature and an ecocentric view of the world as inextricably linked. These preconceptions obscure the possibility that animals can be simultaneously revered and viewed as resources to be exploited (301–5).

One of the most prominent and persistent versions of this commitment to the concept of the ecologically noble Indian is the argument that Native American hunters view animals as "nonhuman persons" with whom they are engaged in cycles of reciprocal exchange. Some scholars are fiercely literal on this point: Timothy Ingold (2000) argues, for instance, that for northern hunters, "animals are not *like* persons, they *are* persons" (51; see also, e.g., Descola 2013).[17] Nadasdy (2007) suggests that Canadian wildlife authorities ought to incorporate "conceptions of animals as persons who engage in reciprocal social relations with human [hunters]" as part of a factual basis for wildlife management decisions.[18] The confusion caused by the use of the word *person* in this context has been seized upon by misguided attempts to justify Makah whaling on the basis of its spiritual significance alone—a version of what Matei Candea (2014) calls "identity politics by ontological means."

I have never heard or read of an instance of a Makah whaler using the word *person* to describe a gray whale, but much of the talk around Makah spirituality could be taken to imply the existence of a thinking, acting subject at the other end of the harpoon (see chapter 3). Where this label fails utterly is in its moral denotation. The Euro-American concept of personhood implies a moral equivalence, including perhaps most notably the right not to be killed for instrumental reasons. Hunting is a relationship of use, and on the issue of killing, at least, it is decidedly *not* reciprocal (it is not expected that animals will ask human hunters for *their* lives).[19]

In addition to moral considerations, there are other reasons to conclude that the use of anthropomorphizing language and kin categories by Native hunters to refer to hunted animals is figurative rather than literal. The first is that Makah interactions with whales display none of the everyday practices and ceremonies that are part and parcel of human personhood in the Makah community. For Makahs who participate in the traditional naming system, the giving and receiving of ancestral names marks the passage between different moments in the life course; whales are given no equivalent. When an activist criticized Makah whalers in the local newspaper for not knowing the whales in their waters because they hadn't given them names, one influential Makah elder responded pithily: "We don't name whales."

The treatment of pets in the Makah community provides a dramatic and useful contrast. Pet ownership in Neah Bay follows a similar pattern to pet ownership in most of the United States (see Tuan 1984). Pets—dogs and cats, mostly, but also rodents and birds—are named, individualized, and given at least some measure of affection. Many are treated like family members, even objects of displaced parental affection. They are not eaten. Micah and Haley McCarty told me a touching story about their first dog, which they adopted as a puppy from among a pack of rez dogs.[20] When they had their first child, the dog went from sleeping on the bed to sleeping on the floor. Then a flea was discovered, and the dog was evicted from the bedroom. "He went from being an only child to stepchild," Micah said, laughing. When the dog died, Micah wrapped him in a Pendleton blanket and buried him across the river. "That's a real, kind of a cultural burial," Micah said.

I do not mean to imply that keeping pets in the American model is the only way to show respect or admiration to animals, or to love them. As a hunter myself, I think it is possible to love and respect an animal and kill it. My reasons for undertaking this project included my interest in hunters' attempts to process the feelings of respect and awe one can have for a hunted animal.[21] Makah pet owners' treatment of their pets, however, is much closer to what Americans (Makahs included) usually mean when they think of *personhood* than is Makahs' treatment of whales. Using the term *personhood* to refer to Makah whalers' relationships with whales is misleading, at best.

Whales, or Whaling?

Some Makah whalers speak in hushed and reverent tones about the secret ceremonies that followed the whale hunt in 1999, and tribal members who were on the beach when the whale was dragged ashore described it as a rapturous experience. But those who would attribute this awe to the sacredness of whales misconstrue the origin of those feelings. Those moments of awe that the whale hunt produced had one thing in common: they celebrated the activity of whaling (and also, in the case of the beach scene, of exercising treaty rights), not whales themselves. Whales are not sacred for Makah whalers, or at least not more than any other animal. *Whaling* is sacred. The images of whales that occur all over Neah Bay support this conclusion. The image of Thunderbird and the whale is not a representation of the whale but rather of a relationship of use between the people and the whale. Makah attitudes toward whales follow much the same pattern as Makah attitudes toward other food animals. Much of what makes whaling different and special, according to the narratives of Makah whalers, concerns the danger of pursuing whales and the traditional and spiritual significance of the activity itself.

There may be disagreement among Makahs over whether whales should be killed and eaten at a given time, but there is rarely disagreement over whether whales are edible. As I have described in this chapter, the prototypical image of "a whale on the beach" reinforces this understanding through its circulation in narratives, tropes, and images. Binki's challenge showed, however, that the edibility of whales is not maintained through those means alone. Her activism ignited a series of reactions that underscore the political and psychological stakes of the moral taxonomy already apparent in the most typical Makah reaction to the mention of gray whales: *yum*.

It is not oversimplifying things to say that the Makah whaling conflict ultimately hinges on a dispute over the edibility of gray whales. However, those who would refer to antiwhaling activism as "culinary imperialism," (e.g., Coté 2010, 204), while they make their point well enough, bury the entire substance of the debate within that witticism. The point of contention between whalers and antiwhaling activists is whether whales *ought* to be killed in order to be eaten. Describing this conflict as *culinary* trivializes its moral dimension. Classifying whales as edible is not simply a matter of taste. This fact is not at all lost on Makahs who say "Yum!" with deliberate conspicuity when asked about gray whales. In categorically opposing the Makah whale hunt, antiwhaling activists were effectively asking Makahs to move whales from the category of things-to-be-eaten to the category of things-not-to-be-eaten—to alter the place of gray whales in what we might call Makahs' moral taxonomy as well as their culinary preferences.

That whales are things-to-be-eaten does not make whaling a mundane activity. The next chapter explores the multiple ways that whales, when they approach the whaling canoe, become entangled in Makah spiritual beliefs and practices, as well as how discourses on spirituality structure the way some Makah whalers *see* whales.

THREE

Everything Is Connected

One evening in early February, I visited Jo Jo McGimpsey's sweat lodge at his invitation. I had met Jo Jo for the first time earlier that day, and he had told me about the "sweat" (as he called the lodge), which doubled as a tribe-sponsored haven for recovering addicts. "I'm an alcoholic," he said. "Been sober for thirty-one years. I guess this is just my contribution." When I arrived, Jo Jo sat reclined in a beach chair near the back of a large wooden deck adjacent to his house. Steam billowed from his skin into the chilly night air. Next to him, a plastic tub overflowed with icy water that was being fed continually by a garden hose hanging over its side. I could smell eucalyptus and woodsmoke, and from inside the sweat came the faint sound of voices. When he saw me, Jo Jo sprang up from his chair and began to show me around, explaining in detail how the sweat lodge operated. As we spoke, a man burst out of the door and leaped over the edge of the plastic tub, plunging cannonball-style into the icy water. Jo Jo saw my look of astonishment and smiled.

It was nearly 250 degrees inside when I timidly entered the sweat lodge a few minutes later. I nearly choked on the dry, scorching air as I fumbled for a place on the bench. There were no lights except for the faint, throbbing glow of the wood stove and the occasional flicker of Jo Jo's flashlight, which was uncommonly dim. There were at least three other people on the benches next to me. One of them was a recovering drug addict who was speaking openly about the substances he used to smoke and the way he used to smoke them. He had been "clean" since the previous fall, he said. He announced it with such a palpable sense of relief that I immediately found myself hoping he could stay that way. Someone sitting nearer to me asked about my project. I did my best to explain what my research was about, but the intense heat was already making my lungs burn and my head buzz. I remember saying something about my interest in the experience of hunting and killing, before I bolted out the door into the waiting relief of the cool night air.

A few minutes later, Jo Jo followed me out. "We were talking about you in there," he said, and I felt a surge of embarrassment for not being able to withstand the heat long enough to hear their replies. But then Jo Jo continued, staring into the blackness past the wooden fence at the back of his yard: "When you go hunting," he said, "you realize that everything is connected. The flowers," he paused, searching for words. "Everything. Everything is connected. And that is why you kill."

A vibrant discourse on "spirituality" surrounds local conversations on Makah whaling, but the topic occupies an uneasy place in the tribe's rhetoric. Some Makahs view spiritual beliefs and practices as appropriate topics of conversation with outsiders. Micah McCarty, for instance, talks about his spiritual beliefs with supporters and skeptics alike, hoping he might change minds or at least make people think twice about opposing Makah whaling. Others, however, feel that the outside world does not need to know about such things. They fear that talk about spirituality will come across as saccharine or sentimental and that members of the whaling crew might be judged on their adherence to some imagined spiritual standard.

The use of the term *spirituality* to describe Native American religious belief has been forcefully critiqued by some Native scholars (see, e.g., Shorter 2012).[1] It is also legitimate to question whether spirituality really ought to be considered separately from other aspects of how Makahs think about whales and whaling. Like so many anthropological terms of art, I find the term helpful as an organizing theme but little else, and I try to limit my use of it accordingly. The word is used frequently enough by Makah whalers and other members of the community; the Makah Whaling Commission charges whaling captains with making sure that crew members are adequately prepared "spiritually and physically," for instance; but it is most often used as an outward-facing term, taking a set of heterogeneous practices and packaging them for easy public consumption.

The topic of spirituality seems to have entered the public discourse around contemporary Makah whaling from several different sources.[2] After their initial conversations with federal representatives in the mid-1990s, tribal officials realized that a commercial whale hunt was not immediately politically viable, but a more self-consciously traditional hunt might be.[3] US fisheries management agencies already permitted Ceremonial and Subsistence (C&S) harvests of commercially important fish species like salmon and halibut, an outcome of the Boldt decision. This provided a convenient bureaucratic category in which to situate a noncommercial whale hunt.[4] The C&S designation also provided a ready-made set of guidelines for the tribe in seeking approval from federal agencies, and it helped them avoid some of the stickier questions associated with commercial whaling, including activists' concerns over the potential for a slippery slope toward renewed commercial whaling worldwide, which was tied to concerns over the possible sale of whale meat to Asian markets.[5]

Makah spokespersons first publicly mentioned the spiritual significance of whaling as early as 1997, and by October of that year, the whaling crew's spiritual preparation had become one of the prominent themes in print media coverage of the hunt (see Gorman 2000). In his 1998 "Makah Manifesto," Keith Johnson explicitly referenced the spiritual importance of the hunt as a defense against outside criticism, writing that the tribe would receive no money from the whale hunt and that the only reward would be "the spiritual satisfaction of hunting and dispatching the whale and bringing it back to our people to be distributed as food." Johnson mentioned that he was initiated into sacred Makah whaling rituals himself as a teenager and that he and others were currently undergoing rituals to prepare themselves for whaling, each according to the traditions of their family lineage. He added, finally, that some of his fellow Makahs would find it repugnant to have to explain any of this to outsiders.

A few Makah representatives have since elaborated on Johnson's sentiment by suggesting that whaling ought to be subject to First Amendment religious protections.[6] "At what point," Micah McCarty asked rhetorically during one of our interviews, "does a philosophical inconvenience become [more important] than the survival of a culture?" Ann Renker compared Makah whaling to Catholic sacraments in conversation with me as well. To prohibit Makahs from whaling, she said, would be akin to prohibiting observant Catholics from taking Communion. In *Spirits of Our Whaling Ancestors*, Nuu-chah-nulth scholar Charlotte Coté observes that antiwhaling activists could only have opposed the hunt by seeing it "through a Western cultural lens" that "ignored the spiritual and sacred elements attached to the Makah and Nuu-chah-nulth whaling tradition" (2010, 164–65).

One important feature of this politicized deployment of Makah spirituality is that it does not depend on much elaboration or detail. There is little incentive to be precise; secrecy brings with it an air of authenticity that seems to only bolster the credibility of a spiritual justification for the hunt. But it would be difficult in any case to describe a Makah spirituality as it pertains to whaling because there is no such thing, at least not in the sense of a unified set of practices or observances shared across all of the extended family lineages within the tribe. Makah spirituality is characterized by its variability.

Some Makah whalers do not engage in spiritual preparation at all. Of those who do, many do so to safeguard against the inherent danger of the activity. A smaller fraction emphasize the kind of relational ontology between hunter and animal that features in many ethnographic descriptions of northern hunting peoples (e.g., Nadasdy 2007). And even among that subset, people vary in how they describe and conceptualize animals' spirits, how they prepare for a hunt, and how they view themselves as part of a broader sacred order. When I refer to Makah spirituality, then, I refer to a heterogenous set of beliefs, practices, and routines that circle a common set of religious motifs without settling on any common configuration. Individuals often piece together their metaphysical understandings from this variegated, proprietary, and secretive material; and the result, predictably, is diversity.

This variation is not necessarily the result of postcolonial fracture and decay (those being inadequate metaphors in any case for describing social change in this setting). There is no formal institution analogous to a church for "traditional" spiritual views, no authority that would normalize spiritual practice or bring its members to heel. Makah spirituality is *normatively* personal and idiosyncratic, and a strong current of local expert opinion suggests that it has always been this way. "In the old days," Greg Colfax told me in one of our many conversations, "religion was a very personal thing. You went into the woods to pray by yourself, and nobody knew how you prayed. If you tried to follow someone into the woods, they would kill you, and they would be justified in doing so."

Today, whaling rituals and prayers are kept secret and usually performed in private. New songs and prayers manifest themselves to individuals—an experience commonly referred to as "receiving" a song—and are thenceforth considered a kind of personal intellectual property when performed publicly. A Makah museum official told me that I shouldn't even ask interviewees about specific prayers and spiritual preparation, since that knowledge was private and not shared even with other members of the community.

This state of affairs—let's call it "normative spiritual individualism"—has a Protestant flavor to it as well, to be sure, and most individuals in the community do belong to one or more of the five Christian churches on the reservation. Much spiritual knowledge, like so many other aspects of social life in Neah Bay, is transmitted along family lines, which helps us to understand why Donnie Swan would say to Robert Sullivan that it was easier for him to work with Wayne Johnson because, as cousins, they "share the same religion" (Sullivan 2000, 192).

The goal of this chapter is *not* to chart this diversity in all its forms, let alone is it to provide an authoritative picture of religious life in Neah Bay, which is far beyond the scope of this book and arguably not my prerogative. The goal of this chapter, rather, is to explore how spiritual ideas and practices serve as optics for engaging with nonhuman animals—incipient forms, perhaps, of what Ingold (2000) calls "skills" for perceiving the environment (5). My aim here is to help clarify how it is that the phrase "Everything is connected" could lead to "That is why you kill." A grounded and participatory spirituality can cross borders between politics and phenomenology, as practices that might have had their origins in a kind of performativity eventually come to structure the way that whalers experience the hunt. This is evident in some of the more intimate spiritual experiences shared with me by members of the 1999 whaling crew, which I recount near the end of this chapter. The picture of Makah spirituality that I present here is, nonetheless, necessarily partial and incomplete, limited by issues of access and by my particular subject position as a white anthropologist and a visitor in the community. It is based on what my informants were willing to share with me. So I advance these observations modestly.[7]

Living in Two Worlds

Strategic essentialist claims about Makah identity and sovereignty make their case with an eye fixed firmly on historical struggles, finding meaning in the sustained mistreatment of Makahs by the settler colonial state. Discussions of Makah spirituality, by contrast, often downplay the historical events and episodes that have impacted local religious practice over the last two centuries. A Makah whaler I spoke with was fond of saying that Makahs have been living the same way since the time of the Romans. "If you want to know what things were like in the time of King Tut," he told a group of visiting students at the Makah museum, "ask us! We were on the other side of the world, but we were living exactly the same way." A partisan reading by Miller (2000) suggests the same, presenting the Makah Tribe as "an excellent example of a culture which preserved its important ancient traditions . . . and withstood the effects of time, assimilation by American society, and outright federal oppression, and then revived its cultural, religious, and legal rights by resuming its traditional whale hunting" (167).

The first few generations of Makahs who were called by that name, including those who signed the Treaty of Neah Bay and their immediate descendants, were subject to over seventy years of sustained assimilation pressure during the era of direct oversight by the US Bureau of Indian Affairs (BIA; approx. 1863–1934), during which outward expression of traditional religiosity was prohibited by federal authorities. Agency schools were commonly run by missionaries who viewed social, economic, and religious assimilation as going hand in hand.[8] This era followed a series of devastating smallpox epidemics that killed around three-quarters of the people living near Cape Flattery. "Seventy percent of our people dropped dead within a matter of weeks," Micah McCarty once said to me, reflecting on the lingering impacts of that terrible episode in the tribe's early history:

> You think of the intellectual property, for a people that have trained for hundreds of generations to carry forth the living breath of your ancestors—oral history—for that to just disappear in the matter of weeks, a lot of information was lost. *But a lot was also retained.* Between oral history that is passed down through certain families, there is enough of that talk and interaction with survivors, and what anthropologists have written down, that you can piece a lot of that together.

Christian missionization in Neah Bay did not end with the decline of direct BIA rule and the transfer of greater self-governance to Indian tribes through the Indian Reorganization Act. Jean Vitalis lamented what she called the fragmentation of spiritual teachings in her own extended family (of which she is the head) in a community that she perceived as not fully Christian and yet not quite returned to traditional religion:

> The concern that I have in the absence of Christianity is a belief system that is strong enough . . . for individuals in my family to hold themselves together when it's necessary. . . . Our spiritual teachings are so fragmented because our culture's fragmented. We're 50–50 sort of so we're not, we're still converting, and there isn't enough, in my mind, speaking as a mother, there isn't enough spiritual acknowledgment done in the way we practice our culture now, we definitely need more. And that is so apparent from the, y'know, if you look at the alcohol [and] drug issues we have in the community.

Reflecting on her decades as a tribal judge, Jean observed that people in the community were commonly confused by treatment methodologies like Alcoholics Anonymous that begin by appealing to a higher power, since "most of the time when our people go they don't know what the hell they're talking about. A lot of them haven't practiced so they don't know about God. But I also know from experience . . . that He's saved some of them, too."

Whether the focus is assimilation or fragmentation or cultural survival ("picking up the pieces"), the above statements might suggest a separation between Christianity and precontact religiosity that has not existed in practice for some time. Today, even those who identify most ardently as traditionalists tend to intersperse their spiritual reflections with reference to a single supreme deity, and those who stress the importance of spiritual preparation for whaling are as likely as anyone to utter a vocal "Amen" after the recitation of Christian scripture at public events. The traditional spirituality of Makah whalers is a Protestant traditional spirituality.

This is not a new state of affairs. Elizabeth Colson, a social anthropologist who sought to understand the effects of the state's assimilation projects on the Makah Tribe through fieldwork she conducted in the 1940s, found that a majority of Makahs approached Christian teachings instrumentally, seeking new sources of what they glossed, in English, as "power" or "medicine," and often switching affiliation between Christian denominations more than once in that pursuit.[9] Colson sensed that most Makah converts to Christianity continued to silently affirm the existence and potency of forest spirits and other traditional sources of spiritual power, even if they viewed their ancestors in a prelapsarian light as the last to summon and wield that power effectively. Even those who spurned Christian churches altogether during Colson's time puckishly identified themselves as "heathens," defining themselves in opposition to Protestant teachings.[10] There were likely some exceptions: the Indian Shaker church forbade ritual songs and dances altogether, and as early as the 1940s the Neah Bay Presbyterian church had an evangelical character that posed some additional challenges to its future adherents when they sought to rediscover traditional Makah songs and dances.[11] According to Colson, there were also Makahs in the 1940s who already viewed themselves as the true and original Christians (1953, 292). This sentiment survives today. "Christianity hit our shores here, it fit us like a hand in the glove," Gary Ray, a politically active woodsman who delivered weekly Christian sermons to his many grandchildren in his home near the Tsoo-Yess River, once told me:

> You can go through principle after principle of Christianity and that's who we are as a tribe. It's more blessed to give than to receive. That's a biblical principle. Rise in the presence of the aged. Show respect for elderly. Deuteronomy. That's a biblical principle. Only be careful, watch yourself closely, do not for-

get the things your eyes have seen. Do not let them slip from your heart, but be careful to teach them to your children, and their children after them. That's a biblical teaching. On and on it goes.

Not everyone in the community agrees, however. Some people are hostile to the presence of Christianity on the reservation and blame Christian churches for the deliberate destruction of Makah traditions. One self-styled traditionalist told me she had refused to accept a local church's offer to build her a smokehouse for smoking fish because she "just couldn't cross over to Christianity." A fisherman told me emphatically that he does not go to a Christian church. "I follow what my grandfathers said," he said. "I think when I die I'll go somewhere."

Judeo-Christian terminology has certainly made its way into the everyday language that Makahs use to talk about human responsibility for the natural environment. It was common in interviews and other settings for Makahs to talk about themselves as "stewards" and caretakers in a way that suggested human dominion over the Earth as given to humans by a creator. Greg Colfax suspects that combinations of Christianity and traditional Makah spirituality often "work together in the minds of the individual who is practicing it" but can be "confusing in the ears of those who are listening to it." His grandparents were Christians who also maintained traditional Makah spiritual practices and attempted to pass them on to their children. "They lived in two worlds," Greg said of his grandparents. "And they lived in two worlds excellently."

Not Getting into the Spiritual Stuff

Any discussion of the spiritual aspect of Makah whaling must acknowledge that some Makah whalers engage in no spiritual practices at all, and in most cases, this does not make them any less *Makah* or exclude them from participating in whale hunts as long as they possess what is perceived to be the requisite hereditary authority. Whaling captain Wayne Johnson made it clear in his comments to Robert Sullivan that he wasn't "too good with the spiritual stuff," and he repeated this sentiment to me on a number of occasions. Johnson received criticism from some members of the community for his comments, but he is far from the only whaler to feel this way. Several individuals associated with the 1999 hunt did not engage in spiritual practices or felt that spiritual considerations should not have any impact on the planning and implementation of the hunt. The fact that Wayne, as the whaling captain, was charged by the Makah Whaling Commission with verifying the spiritual readiness of the crew created no small number of difficulties for him, which Robert Sullivan narrated with requisite irony in his account of the hunt (2000).

It would be unfair to portray Wayne in this light as simply *lacking*, however. He clearly finds fulfillment and takes pride in the technical knowledge and mastery associated with whaling as well as in his ability to read the movements of whales and anticipate their behaviors. "There's only a few guys in Neah Bay that can get a whale and bring everyone home safely," he told Linda Mapes of the *Seattle Times* after the rogue hunt in 2007. Wayne has traveled as far as the Russian Arctic on tribe-sponsored trips to learn new techniques from other whalers trained in the Norwegian method of small-boat shore whaling, and he was eager to demonstrate that knowledge whenever the topic of whaling came up, which, at least in our conversations, was often.

Preparing for Danger

For some Makah whalers, the main object of ritual preparation isn't the relationship with the whales at all but is instead the safety and effectiveness of the hunt. Just as some Makah fishermen pray to ancestors for a good catch and for a safe return, these whalers prepare themselves physically and spiritually to ensure their safety and success in pursuing such a large and powerful being. "A whale can kill ya a lot faster than an elk can," one whaler told me when I asked why spiritual preparation for whaling was important. "Plop, you're dead. The whole crew's gone. Y'know you gotta approach things a lot different." Some of these individuals expressly reject the notion that whales might offer themselves to humans, finding the idea a little outlandish. The aim of their spiritual preparation is to safeguard against the inherent dangers of hunting a large and dangerous animal from a relatively small and fragile canoe. While this form of spirituality includes similar practices to the relational spiritual ontology we will talk about in a moment, including bathing in cold creeks, flogging oneself with stinging nettles, and so on, the aim here is strictly inward facing, a cultivation of discipline and prowess that will help ensure the success and survival of the hunter.

Asking the Animal for Its Life

For some other Makah whalers, ritual preparation happens in the context of a relationship between the hunter and the animal spirit. This is the viewpoint that has most captured the attention of scholars working with North

American indigenous societies (e.g., Ingold 1987; Brightman 1993; Nadasdy 2007) and that has been held up more recently as emblematic of the "animist" viewpoint within anthropology's ontological turn (Descola 2013). In the context of such relationships, the spirit of the animal is said to respond to the spiritual preparation of the hunter. If a hunter does not prepare properly, the animal may not offer itself. Spiritual preparation in this case becomes an index of worthiness, of a kind of respect for the whale, and an indication to the whale that if it offers its life to the hunter its spirit will be properly "taken care of." This way of thinking is common enough among Makah whalers today that for many Makahs, the phrase "A whale gave its life" has become synonymous with "There was a successful whale hunt."

Relational spiritualists believe that if the hunter prepares himself properly (e.g., by bathing, praying, abstaining from sex, or other secret rituals) the whale will acquiesce and "come home," offering its life to the hunter to feed his family and the village. This inference is usually made after the fact. I was told by a number of people that in the old days, if a hunt was unsuccessful, people in the canoe would turn and look at each other and say "OK, who slept with his woman last night?" Not only does the whale's spirit know if the hunter has prepared properly, it is also risky to even talk about going whaling in the days leading up to a hunt. Hunters avoid referring to the whale by name and instead speak about it indirectly.

Ideas about the whale's predisposition to be flattered, cajoled, and induced to offer itself to whalers recur in whaling songs passed down through generations, as in the Nootkan whaling prayer recorded by Edward Curtis:

> Whale, I have given you what you wish to get—my good harpoon. And now you have it. Please hold it with your strong hands. Do not let go. Whale, turn toward the fine beach of Yahksis, and you will be proud to see the young men come down on the fine sandy beach of my village at Yahksis to see you; and the young men will say to one another: "What a great whale he is! What a fat whale he is! What a strong whale he is!" . . . Whale, do not turn outward, but hug the shore, and tow me to the beach of my village of Yahksis, for when you come ashore there, young men will cover your great body with bluebill duck feathers, and with the down of the great eagle, the chief of all birds. . . . (Curtis 1911, 23; reproduced in Waterman 1920, 39)

The whale's telos—what it "wishes to get," as the song puts it—is to die at the hand of the worthy hunter. The inducements offered to its spirit are things that are of value from the hunter's perspective, such as flattery and eagle down.

These themes of reciprocity, flattery, and gratitude, along with the imputation of a kind of volition on the part of the animal spirit, have led a number of scholars to argue that people in North American hunting societies view whales and other animals as "nonhuman persons." I pointed out what I believe to be the inadequacy of this conceptualization in the previous chapter, particularly with respect to the moral dimension of the hunter-whale relationship. The concept of nonhuman personhood also inadequately represents what we might call the spiritual relationship between hunter and whale. Here is how Micah McCarty described the importance of spiritual preparation for whaling:

> It boils down to your own individual responsibility, if you take the life of such a magnificent creature, that you prepare yourself mentally, physically, and spiritually. You take that responsibility of ushering that spirit, caring for that spirit, and preparing it to leave its body. . . . My great-grandfather's Indian name was Hishka, and basically what that name means is "he makes the whale blow on the beach." That is an old, old ancestor, one that was able to do such a thing as have a whale tow him home. And what happens after the whale is blowing on the beach is a ceremony to usher the spirit of the whale into the next world, in a manner in which it's accepted by those who came before it, in the same manner. So that connotes a society of whales that have been hunted, in the spirit world, and they welcome the new ones as they are ushered into the next world.

The image of a "society of whales" that have been hunted for generations naturalizes Makah hunters' place in the spiritual order, running in parallel to a key component of Micah's political rhetoric in which he positions the Makah Tribe as a "keystone species" without whose predatory activities the local ecosystem would collapse.[12] Both moves inject a deep temporality into the discussion. The whale's approach is no longer situated in the context of a particular hunt or in a relationship with an individual hunter but in a spiritual history of hunters responsibly harvesting whales and ushering their spirits into the next world. This has been going on for generations on each side, human and whale. Such a vision casts whales as possessing individual spirits that move into the spirit realm. Others in the community sometimes speak of "the whale" as if it were a single metaphysical being, however, as if each individual animal were an instantiation of this transcendent spirit. The latter is how animals tend to appear in recorded myths, with the trope of a "council of animal spirits" at the beginning of time made up not of representatives of species but of "the whale" and "the wolf" and so on.[13]

One thing I want to emphasize from Micah's comments is how much they are reminiscent of another category of relating to nonhumans: care. Taking care and taking responsibility for the whale's spirit are recurring motifs in reflections on the importance of spiritual preparation offered by Micah and others with similar perspectives. This is by far the most common way one hears people talking about a whale in the context of spirituality, taking care of the whale. It is important to note that care does not entail the personhood or even the sentience of the object in question. Anthropologists have shown how one can exercise care over inanimate as well as animate objects. Like Micah's parallel tropes of a society of whales and an ecology that includes Makah hunters as keystone species, care unites material and metaphysical concerns about gray whales that environmentally conscientious Makah whalers and Makah officials prioritize. The romanticism of the ecologically noble Indian may be thoroughly rejected by Makah hunters and administrators alike, but the idea that Native American hunters are the wise and original stewards of natural resources has been accepted and promoted by many people in Native American communities across the American West and Northwest.[14]

Spirituality and Technology

Contemporary Makah whaling takes place in a modern context, with modern implements and within a rationalized bureaucratic environment designed for the management of tribal commercial fisheries. Current and former commercial fishermen were heavily involved in the planning and implementation of the 1999 hunt as well, and Makah commercial fishermen have a reputation for expediency that would seem to clash with the slow, deliberate reflection that spiritual preparation and practice call for.[15] A desire for expediency is rather understandable given the sheer volume of animal life that fishermen deal with on a day-to-day basis. It is easier to imagine someone engaging spiritually with a single large creature at the center of a hunt years in the making (like a whale) than to imagine fishermen maintaining that kind of immersive spiritualism for an entire, grueling shift. Micah McCarty has thought deeply about this issue, and he struggled with it as a commercial fisherman and deckhand:

> The way you clean a black cod is you grab it by the head, you flop it onto its back, and then in two cuts you cut the collar and head right off and then throw the head over[board]. Grab the belly wall, pull out whatever remaining guts, hose out the blood line and then throw that carcass into the ice hold. And I

can't help but to acknowledge I just killed something. I just killed a fish that's beautiful and sleek and tasty. And it was alive until I cut its head off, and I couldn't help but to think what that fish must be thinking with its bodiless head sinking down into the ocean. That it must still have a little flicker of awareness. Because its head was just severed, within seconds it's overboard, and the body's still kicking. And that's on the back of a fishing boat.

He looked up from the paddle he was carving as we spoke. "And I still have that feeling."

"Did you have ceremonies in the fishing boat [for the fish]?" I asked.

"Inside my head!" he replied. "[But] when you're dealing with hundreds of fish, [you] can't do a ceremony for every one of them."

Those Makah fishermen who do not focus on a relationship between hunter and hunted struggle less with this conundrum. Rather than thanking individual fish, Chad Bowechop focuses on the ecosystem at large in his ongoing attempt to incorporate spiritual teachings into modern resource management.

"I didn't waste a second trying to communicate spiritually with that salmon, I guarantee that," Chad said, reflecting on his experience as a fisherman. "I was grateful for [the salmon's] place in the ecosystem . . . [and that] the ecosystem was healthy."

Chad describes the ocean itself as not just a physical but also a spiritual medium. Drawing on his father's teachings, he feels that the pollution of the ocean creates "background static" that interferes with how people today can understand and communicate spiritually within that environment. Still, Chad feels that the adoption of larger ships represents an "interruption," since he views spiritual and physical preparation as a response to the conditions (and hazards) of the hunt.

Questions about the tensions between tradition and technology and between spirituality and expediency were at the heart of debates over the form the 1999 whale hunt would take. Early Makah boosters of a return to whaling envisioned the use of motorized boats or even larger fishing vessels, especially during the winter months when the seas would be too rough for a small canoe. Proponents of the higher-tech approach pointed to the tribe's long history of adopting new technology when it offered clear advantages, from rubber buoys and metal harpoon points to underwater sonar and large fishing vessels. Makah seal hunters used large ships to pursue fur seals for sale in commercial markets as early as the nineteenth century, and they had petitioned the federal government for assistance in buying modern steam-powered whaling ships during that era as well.[16] One Makah man I spoke with was particularly annoyed by the use of paddles in the contemporary whale hunt, pointing out that Makahs had adopted oar technology very early in their postcontact history, allowing them to achieve much faster speeds with their canoes. It did not make sense to him to go back to the inferior technology of paddles. Paddling has become so integral to the resurgent indigenous canoeing culture on the Northwest Coast, however, that I do not expect Makah whalers will use oars any time soon—in fact, the annual canoe pilgrimage known as Tribal Journeys is usually subtitled "Paddle to [the host city]."

Sometime before the 1999 hunt, the Makah Whaling Commission—composed of representatives from the twenty-three whaling families—decided it would be best if the hunt were done in a more traditional manner, using a dugout canoe and a nonexplosive, or "cold," harpoon. Planners added a small motorized chase boat in order to appease US Coast Guard concerns over the safety of the hunt. At the behest of the IWC, the Makah Whaling Commission agreed to require the use of a firearm in the hunt to make the kill faster and more humane.[17] A canoe-based harpooner accompanied by a motorized chase boat carrying a shooter armed with a high-powered rifle emerged as the go-to configuration for Makah whale hunts from 1998 to 2000, such that when the whalers in the 2007 rogue hunt used only motorized boats, it added to the impression that the hunt was illicit and rushed.

Some Makahs felt that traditional methods would be intrinsically valuable regardless of whether the hunt was associated with spiritual beliefs. Those who felt that whaling could help solve the community's drug problem or improve the discipline and self-esteem of troubled youth fall into this category. Janine Bowechop pointed out that the rigors of physical training, or "getting to the point where you can paddle for miles and then maneuver where you need to be right next to that whale," were good for promoting fitness, strength, and endurance among young men. When Darrell Markishtum reflected on the hunt years later, it was clear how deeply he missed the camaraderie of the crew. "We were closer than family," he said—not at all an insignificant comment to make in a community where family accounts for so many social bonds—"closer than brothers." Returning to "normal life" after the hunt, he said, was "pure torture." Darrell later served as a probation officer in the area, and he considered getting the guys who were on probation into a canoe and putting them through whaler training as part of their rehabilitation.

Not everyone agrees that traditional whaling methods (most notably, the use of a dugout canoe) are necessarily bound up with spiritual preparation for whaling. Some believe that prayer, preparation, and the relationship between whaler and whale are important and valuable but not dependent on a rough equivalence or level playing field. In describing his experience butchering black cod, Micah points to a way that spiritual engagement might continue in a more industrial setting. He once suggested to me that it would be relatively easy to have a ceremony for the whale on the deck of a commercial fishing boat. Dave Sones envisioned two kinds of whale hunts in the future: a traditional, or cultural, hunt that would receive the first quota, followed by a commercial hunt that would start after the first whale had been brought to the beach. Sones felt it would be nice if they brought in at least one whale per year with a canoe and cold harpoon, since he recalls that doing so in 1999 was a "great experience." Theron Parker agreed: "I always thought the big ship thing was a foolish gesture unless you're gonna do commercial whaling," he said. "The safety of the crew, that was an issue, but I always said, let's do it that way [with a canoe]. The first one, to show that we can, to show we are who we say we are, let's do the first one by canoe. There was a lot of controversy about that. I said well, I don't care, you guys wanna take a ship, go ahead. I'm still gonna use a canoe."

Spirituality as Skill

In discussing the various ways that people perceive their environment across social groups, Timothy Ingold refers to the "skills, sensitivities, and orientations" that emerge through long experience. Ingold stresses that *"skills* are not transmitted from generation to generation but are regrown in each, incorporated in the *modus operandi* . . . through training and experience" (5; first emphasis added). One thing to like about Ingold's formulation is that it makes room for what many Makah whalers feel is the rediscovery of religious practices through the revitalization of the traditional culture around whaling. Routines and practices passed down through generations can feel as if they exert an influence from outside the individual. This kind of experience is, in part, what led Theron Parker, the harpooner of the 1999 hunt, to tell me in an interview years later that he felt he was "just a tool" for the spirits who acted through him.

In the months leading up to the 1999 hunt, the Makah Whaling Commission effectively left physical and spiritual preparation to the individual prerogatives of the would-be members of the whaling crew, each according to his family's traditions.[18] Spiritual advisors like Greig Arnold and Jo Jo McGimpsey helped prepare some members of the crew spiritually, offering advice and spiritual guidance. After the hunt, Theron Parker singled out Jo Jo and Greig Arnold with gifts of the finest pieces of the whale. "I was fortunate to help prepare the whalers spiritually," Jo Jo said, before telling me a humorous story about how he had not realized the piece was for him and had unknowingly delivered it with the rest of the meat to a communal storage facility. "That's okay," he said. "More people got to enjoy it."

Wayne Johnson absorbed much of the media attention before and after the 1999 hunt, but Theron Parker became the face of whaling in Neah Bay, a popular figure whom many credit with bringing the hunt to its successful conclusion. "Once Theron got involved," one Makah woman told me, "that's when I knew we'd get a whale." Theron was one of the most serious and imposing figures I met during my fieldwork. When he said "we" to refer to Makahs and "you guys" to refer to everyone else (including me), he said it in a way that somehow made those distinctions feel real and obvious and permanent. A photo of his paternal grandfather, Wilson Parker, posing with a harpoon and whaling buoy at Wa'atch beach is among the most iconic images of Makah whaling in circulation today. "I come from whalers," he told me as we looked at a copy of the photograph in his home in Neah Bay. "My brother and I, we come from the male side all the way down. So we're kinda—we didn't know it at the time, and I don't really talk about it, but we're pretty important people, in our family." Theron showed me a very old harpoon rope that had been passed down to him through generations. I had never seen anything like it outside of a museum.

Theron was far from the presumptive choice to be harpooner when the 1998 crew was first assembled, however. He watched the early stages of the conflict from a prison cell, where he was serving a sentence for a felony drug charge. "I had trouble with the law a lot," he said. "Meth troubles." He credits the annual Tribal Canoe Journeys, an event where tribes from across the Pacific Northwest paddle to a host destination for several days of festivities, with helping him sober up. The event is expressly drug- and alcohol-free and can involve up to a month of paddling, depending on the destination.

Theron remembers seeing a TV news feature on the Makah hunt while still in prison and telling a friend, "I'm gonna go do that. That's where I'm from." When he returned to Neah Bay, he and a couple of guys watched the putative Makah whaling crew from the beach during a prearranged photo shoot and were unimpressed:

> We'd call them the "Hollywood whalers," they'd oil their bodies up and perform all these fucking glamorous little things that they would be doing, y'know? We're sitting here watching this shit on the beach, like, look at these guys, man, what the hell? They're not getting ready. Every time a camera shows

up they go get in the canoe and pretend like they've been practicin' and pullin' and workin'. We were like, fuck it, let's just go do it—this is my language, this is how I talk, this is how we are, anyway I'm just a guy, y'know, what the hell— but anyway, so, we're watching this stuff, and every time a camera shows up, they'd be looking for the canoe, they'd jump in there, they'd oil themselves up, they'd act like they're all this, and like, they ain't gonna get shit, they're just, they're just, *performers*. We thought it was a joke.

Theron clearly has little patience or interest in public-facing aspects of the whale hunt. He told me his interview with me was the first he had granted in years, and he had done so primarily to air grievances over the tribal government's handling of the 2007 rogue hunt. Robert Sullivan had noticed that Theron "didn't show up in the papers or on TV at all when they were covering the whale hunt; he didn't seem to like most of the reporters, much less the idea of doing interviews" (2000, 192). Theron viewed an excessive focus on publicity as antithetical to success in the activity, especially if it came at the expense of actual training. He had never hunted a whale before, but he had always hunted deer, elk, and other animals, and he knew that killing such a large animal from a canoe using only paddles for propulsion would require a great deal of preparation. His metalinguistic aside in the above quotation ("this is how I talk . . . I'm just a guy") further distances himself not only from those tribal members who were deemed articulate and camera ready and upstanding enough to represent the tribe in the whaling canoe but also from those who were invested in whaling as a political project.

"They weren't the top warriors in our village," Gary Ray once said to me of Theron and several others who ended up on the whaling crew. "They were the throwaways! You know, the druggies, the guys who didn't really have anything, [but] they chose to step up to the challenge." John McCarty was more blunt: "Basically, they're fuck-ups," he told Sullivan in 1998 (Sullivan 2000, 215), "and I'm getting tired of lying for them."[19]

Whether or not Theron views himself in such stark terms, it was clear that he began on the outside of the whaling effort. His solution was to say, "Fuck it, let's just go do it," to let the strength of his commitment and his reputation as a serious-minded person in the community supersede the official pronouncements of the tribal government. He and a few other men started taking the whaling canoe out at three o'clock each morning, before the official crew was awake. "We'd go paddling all over the place," he said, "[and] have it back by six, seven, eight. Before they even get out of bed, we'd be done. They'd come down and wonder why the canoe was all wet."

Micah McCarty eventually stepped aside from the harpooner role, as did Eric Johnson, and Theron took their place. It was around this time that Theron and his close associates on the crew quit their primary jobs and dedicated themselves entirely to preparing for the hunt, with Theron taking the initiative in bringing crew members to the museum to learn about whaling in the old days.

"After a while it become just straight getting ready. We got a sweat house, did a lot of praying in there. Had a lot of songs that come to our heads, songs from who knows where. We got those, and we go bathing in creeks. I started bringing a couple guys that were really interested up to the museum, researching history, talking to all our old folks that were alive at the time [of the last whale hunt]."

Theron also took over the task of maintaining the *Hummingbird*, the cedar dugout canoe that the crew used for practice and that would eventually be used in the 1999 whale hunt. What finally caused Theron to commit himself completely to his physical and spiritual preparation for the hunt, however, was his crew's first experience with a gray whale from the canoe. "The first time we went out, [a whale] came up, its eye was like right above the water. I was lookin' at him, he was lookin' at me, and he just [sunk] down slow and just flicked his tail and *whoosh*, right by us, real fast. I said 'Fuck! We are so outclassed, we don't even belong out here right now.' I told the guys to pack it up. 'We are so outclassed. We are *so* outclassed. We are not even ready to be here. We cannot do this. We cannot do this today.'"

His solution, he told me, was more *everything*—more spiritual preparation, more strength training, more of what he called "dialing in." "We gave up everything," he reiterated. "*Everything*. Sleeping with our ladies, we didn't work anymore, people would feed us. I don't know how we got the bills paid," he said, looking back. "I really don't know how it happened. I don't know. Because I didn't even care about that. None of that shit even mattered. Nothing mattered. . . . I didn't think about anything else. I was so far into what I was doing that I didn't even think about the other side of the world."

It seemed to Theron that the impetus for everything important happening in his life was coming from outside himself, and this was literally true in a secular sense as well. "It takes a village," Theron said, reflecting on what he learned about whaling. The first executive director of the Makah Whaling Commission, John McCarty, had tried and failed to procure grant funding to support the original whaling crew during their preparation. Theron and his fellow crew members accomplished the same goal through informal social networks in Neah Bay.

When Theron told me that he was "just a tool" for the spirits that acted through him, I at first thought that his comment was part of a spiritual motif, a bit of discourse that he had picked up while learning about the old ways. I later realized how that feeling might have arisen directly from his experiences, from the whale's prodigious speed, which led Theron and the others to prepare so rigorously, to the support of the community, which he felt enabled his preparation. Theron recalled something an elder had said to him when he was still a young man, long before the tribe considered a return to whaling. "A long time ago, when we were whalers," the elder had told him, "the men that were getting prepared—at every body of water they would stop and pray."

"I didn't think nothin' of it," Theron said, "until I started getting ready [for the 1999 hunt]. Every body of water you come across, you pray. So pretty much you're just prayin' all day long. *All day long.* You can't stop, it just becomes a thing. Toward the end of it, everybody said we had this look about us, this aura that was frightening. Everybody would back away, they wouldn't talk to us or come near us."

Ritual observance, in other words, changed Theron in ways he did not anticipate, his very perception of the world reconditioned by the observance of a practice hundreds of years old and passed down to him. Theron searched for words to describe how it felt to actually go on a whale hunt in this excited state, when a whale is finally sighted. "If you've ever been on a roller coaster . . . that pounding feeling in your heart, how you're just like super-amped, that's how you feel. You're so alive, your ears are more crisp than they've ever been, your taste is more tastier, the smell, your eyes, everything is like so, so intensely focused. Like if you were newborn, experiencing life for the first time, on top of having like eight Red Bull [energy drinks] inside of you."

For all the heights Theron experienced in the course of the hunt, he was careful to stress the limitations of his knowledge compared to people in the old days. "We're nine-to-fivers," he said. "We don't understand, because like when you're back in the [old] days, you're thinking about how the hell am I gonna get something to eat? There's a lot to eat right there [in a whale]. How am I gonna get that? So that's all you gotta do all day long. . . . Like us, we gotta worry about, well, is the heat bill due today, do I got a car payment tomorrow? You can go to McDonald's to get something to eat."

Spiritual Epistemology

Theron's spirituality is strikingly ecumenical. Sweat lodges are borrowed from Plains Indians, along with the use of sage ceremonies. Theron relied on several people from Nuu-chah-nulth tribes to help him prepare spiritually and others for important practical knowledge on preparation for a whale hunt. Jo Jo McGimpsey similarly told me he learned much of what he knows from spiritual mentors from other tribes, including Nuu-chah-nulth elders. Theron utilized written as well as oral histories, along with ethnographies from the Makah Cultural and Research Center's collection. Robert Sullivan observed that Wayne Johnson kept the seating chart from T. T. Waterman's *Whaling Equipment of the Makah Indians* on the dashboard of his van like a road map during the lead-up to the 1999 hunt (2000, 201). It was the same volume that Greg Colfax consulted when Dan Greene asked him to help butcher the whale he had caught in his set-net.[20] In discussing the fragments of Makah spirituality that have survived settler colonialism, Micah McCarty mentioned classic ethnographies in the same breath as oral histories.

All of this is characteristic of the multiple paths that the transmission and diffusion of spiritual knowledge and practices have taken for Makahs. For many people in the community, their first exposure to the idea of spiritual preparation for whaling comes through the ethnographic work of Edward Curtis, James Swan, and T. T. Waterman, all of whom recorded rich descriptions that the whalers of today find useful. Waterman's account (1920) is based on interviews with a member of the Peterson family but seems to have a broad appeal nonetheless.

The "experts" on Makah spirituality to whom I was continually directed make little distinction between these two epistemological forms, that is, between knowledge that has been passed down through generations and knowledge garnered from reading or hearing about early ethnographic accounts. Both are considered authoritative and valuable. Darrell Markishtum, a deeply religious member of the 1999 crew whom almost everyone I spoke with identified as a good person to talk to about spiritual beliefs, once told me a story about a whaler who was caught stealing a skeleton from a graveyard to use in a whaling ritual. It was his brother's daughter's remains, and when the whaler was caught in the act, he "dropped dead of shame." When Darrell finished the story, I asked him where he had first heard it, and he replied that his grandpa had told it to him, and it was in "one of the books."

Greg Colfax recalled a conversation he had had with a Makah man who had spent hundreds of hours trying to reverse-engineer a kind of traditional nettle-fiber fishing line that had not been used for generations. "I tried to hand him the book that would tell him how to do it," Greg once said to me, feigning exasperation. "He said he wanted to learn it himself!"

Everything Is Connected

One cool evening in May, I sat and talked with Greg Colfax underneath a canopy he had stretched over a hand-carved house pole in his backyard. Years ago, a famous movie producer had commissioned Greg to renovate the pole but had subsequently gone bankrupt, leaving the pole in Greg's possession. Greg was thinking about finishing it in order to recoup some of the value of the time he had invested in it. "I need money!" he said, his eyebrows arched upward in a self-effacing grin. We sat together until the sun had set and the twilight gathered around us, with him puffing on his pipe and me digging through a bag of gluten-free tortilla chips that his wife, Linda, had brought out for us to munch on. I talked with Greg about his childhood in Italy and New Jersey under the care of his Florentine mother, a line of questioning that led him to ask me about the purpose and methods of psychological anthropology. We talked about his son's fledgling efforts to captain his own fishing boat. "He's an excellent fisherman," Greg said of his son. "His heart is all there. But he's a guitar player. So, the question is, can a guitar player become a boat mechanic?" We talked about Greg's interest in comparative religion, and Makah whaling, and the ritual practice of whalers' wives.

"There is a part that you probably won't get," he said finally as he stared into the middle distance. "Everything I've read so far, Boas, others, they didn't get it. If people aren't going to believe, for instance, that the [whaler's wife] can almost become the whale in this act of prayer, then there's no point going any further with them. If their culture won't meet us halfway, then I'm not going to wait for them. . . . There is no knowledge without suffering. Your leaders haven't suffered enough to be in the room with us. This discourages me."

Greg's grandfather, Skyler Colfax, converted enthusiastically to the Indian Shaker church well before Greg was born and decided to cease all traditional spiritual practices and sever all ties to his family's religious past. Greg's family does not talk of whaling gear or the accoutrements of traditional status, and none of his family members joined the whaling crew. He has indulged a keen interest in the topic of traditional Makah spirituality, however, and his erudition—along with his prodigious talent as a carver of ceremonial masks—has earned him a local reputation as a bearer of cultural expertise.

"I've tried to read every possible thing that Native whaling men and their wives have said about whaling," he told me. "So, I have a sense of a little bit about what it was like long ago."

Some of what whaler's wives have said was recorded by anthropologist Erna Gunther in her 1942 essay:

> After the canoes had gone the whaler's wife returned to her house and lay down on her bed (in a darkened room after they moved in the white man's type of house). There she remained without moving until about two o'clock the following afternoon. It was during this period that the hunters might encounter a whale. If the wife moved about, the whale would be unruly and hard to catch. She was told not to eat or drink anything until she rose, when a small amount of dried fish could be eaten, but nothing more.
>
> [One whaler's] widow tells of her first experience as a whaler's wife: "When I got up about two o'clock I ate a little dried salmon and I was so thirsty that I drank some tea. When my husband came back he walked up to me and said, 'You drank something when you got up; we got a whale but he is not fat.'"

During the 1999 hunt, several of the whalers' wives and girlfriends chose to revive this spiritual tradition. "The whole hunting event becomes a reenactment of the husband and wife's physical love for each other," Greg explained to me. "The whole thing comes together. It's all connected. The wife has to bury herself in the corner of her house. She becomes so much like the whale that there's a danger of her *becoming* a whale."

"Everything is connected." This phrase serves as a motif among the spiritual leaders with whom I regularly spoke, a basic truth from which is derived the social fact of the hunt itself. "Once you have it in your mind how you're gonna do it, how you're gonna go and hunt a whale," Greg told me, "once you have it in your mind, all the training and all the practice, and then the leaving of the beach and meeting up with the whale that's coming to greet you, coming to meet you, coming to give its life to you, then you're in another realm. Then you're in, you are participating in your life's philosophy. The world is all one. That is the ground floor of our belief system."

FOUR

This Fishery Will Be Managed

If given a god's-eye view of the first Makah whale hunts in the fall of 1998 and spring of 1999, one would not have needed to scan the scene long to find indications of the US federal government's influence. The National Marine Fisheries Service (NMFS) was directly involved in authorizing and planning the hunts, contributing stock assessments, implementing federal and international regulations, and helping the Makah Whaling Commission draft its management plan. The Makah whaling crew was not permitted to hunt at all unless a federal observer was present. NMFS usually had its own vessel on the scene at all times for this purpose. When the whaling crew wanted to begin hunting, they would hoist a flag from the stern of their canoe, at which point the US Coast Guard would begin enforcing a 500-yard "moving exclusionary zone" around the canoe.

Makah officials have publicly maintained that their treaty-backed right to hunt gray whales is exempt from all federal regulation and oversight,[1] but tribal leaders never seriously considered excluding NMFS from the planning and implementation of the whale hunt. The tribe's return to whaling was first championed by tribal fishermen who had extensive experience working within the established channels of fisheries co-management. Federal legitimacy for whaling was part of their vision. In the years leading up to the first hunt, the tribal government committed several full-time employees to the task of navigating the federal bureaucracy associated with whaling in the age of the Endangered Species Act (ESA) and Marine Mammal Protection Act (MMPA), a task that Makah officials referred to as harvest management.[2] It was over the winter of 1998, Robert Sullivan wrote, that "the sacred act of whaling became a kind of bureaucracy in itself."

Makah leaders won federal regulatory approval for a whale hunt thanks in part to their comfort and familiarity with the federal fisheries management system. According to a number of Makahs with whom I spoke, this comfort indicated more than just good working relationships between Makah officials and federal agencies. They saw a deeper affinity at work, one based on the overlap between the traditional allocation of limited seasonal resources by Makah heads of households and the demands of modern bureaucratic conservation. This chapter explores that affinity, relying mostly on the explanations of Makah administrators, fishermen, and whalers. The chapter concludes by considering the moral consonance between the two discourses, which, I argue, enables Makahs to successfully translate their whaling agenda into the language and logics of fisheries management.

A Functioning Element of the Ecosystem

Technocratic authority is based on appealing to epistemological standards that exclude rival systems of knowledge.[3] For modern fisheries management, this has customarily meant privileging policy-relevant science based on systematic observation and measurement over lay or traditional forms of knowledge, which appear from such a vantage to be partial, biased, and unsystematic.[4] In recent years, however, efforts at recognizing "traditional ecological knowledge" have begun to take into account alternative forms of ecological knowing, including bodies of inherited knowledge about ecosystems and their functional interrelationships.[5]

Reducing nature and society into simplified and legible symbolic forms has been a characteristic feature of the modern state; and, as Patchen Markell (2003) observes, this mandate has repeatedly tied the project of state sovereignty to efforts at engineering cultural homogeneity (152).[6] For this and other reasons, anthropologists have been skeptical of whether projects that seek to incorporate traditional ecological knowledge into the decision-making process of the modern state can ever be salutary to the interests of indigenous peoples. Some scholars have lauded instances of cooperation where Native American governmental institutions effectively utilized biological science that was "enhanced and complemented" with local knowledge.[7] Even the most optimistic accounts, however, stress the strategic character of these collaborations, noting the inherent power imbalances that in many cases make state recognition the only alternative to the outright rejection of local or indigenous perspectives.[8] Other scholars have been less optimistic, arguing that modern managerial frameworks will invariably delegitimize the very indigenous knowledge they purport to amplify. Paul Nadasdy (2007) has asserted, for instance, that bureaucratic wildlife management undermines the political authority of indigenous hunters in the Canadian subarctic, in part by refusing to take spiritual explanations of animal behavior seriously. Elizabeth Povinelli (1995) similarly argues that liberal multicultural attempts to incorporate indigenous traditions and knowledge in arbitrating aboriginal land claims on the northwest coast of Australia actually "entrench state rights over indigenous communities" despite utilizing economic categories that are intended to be emancipatory (516). A strong version of this critique comes from indigenous scholars of postmodernity like Audra Simpson and Glen Coulthard, who argue that recognition amounts to little more than a less corporeally violent form of dispossession continuous in spirit with previous attempts at addressing the "Indian problem."[9]

If I were to report a contrary sense among my informants, it would not necessarily disprove these theories. Nadasdy calls the effect in question "insidious," after all, and members of dominated groups are quite capable of collaborating in the production of discourses that are contrary to their interests. But it is worth emphasizing how starkly the more critical pictures of traditional ecological knowledge and the effects of state recognition diverge from the attitudes manifest in daily life as they were shared with me by people from a broad cross-section of the Makah community. Concepts drawn from fisheries management are widely accepted and used in daily life in Neah Bay, especially but not exclusively by administrators and elected tribal officials. Salient concerns with minimizing bycatch, establishing seasons, and monitoring the population health of targeted species seem to be shared by many who fish or hunt or gather shellfish. These practices coincide with the cultivated sense that Makahs are the best available stewards of local marine resources, an ethic expressed in the common local aphorism "Take only what you need" (see Renker 1997). I was astonished by how often technical terms like "usual and accustomed fishing grounds" (an important legal term in tribal fisheries management dating back to the 1974 *United States v. Washington*) and other bits of fisheries jargon show up in the everyday lexicon of nonfishermen, as when one elder criticized the tribal government on Facebook for allowing a tribal member to shoot a bear near a new housing development. "Bears, cougars, coyotes, and mice are the ones paying for it," she wrote, "because we are [moving into] their usual and accustomed living and hunting areas." The use of such bureaucratic terminology is especially pronounced among whalers, most of whom have experience as fishermen, fisheries managers, or both, and for whom a sophisticated understanding of the bureaucratic machinery of the state seems to be a prerequisite to being taken seriously in the informal exchange of ideas that is constantly taking place at the restaurant or the veterans center or other gathering places.

Perhaps the most widely repeated trope that was invoked by tribal officials in discussions of wildlife during the time of my fieldwork was that Makah hunters were a "keystone species" in the region and that their traditional hunting of seals and other marine mammals had been a vital part of the local ecosystem since time immemorial. Ecologists define a keystone species as an organism that has a disproportionate impact on an ecosystem given its relative biomass (Paine 1969).[10] Micah McCarty, whom I suspected to be the originator of the "keystone species" trope, was acquainted with Robert Paine, the ecologist who coined the term.

Although Micah was originally slated to be the harpooner for the first Makah hunt, he left the crew to attend college. Now in his forties, he has served two separate terms as chairman of the Makah Tribal Council. Micah is articulate and willing to speak openly about his spiritual beliefs, which, in the oversaturated media coverage leading up to the 1999 hunt, made him a media darling. "Yeah, I got recognized by a lot of journalists as a kind of a quote machine," he once said to me.

Micah explained the keystone species argument to me in more detail one day as we chatted in his wood-carving shop on the banks of the Tsoo-Yess River. "We are a functioning element of the ecosystem," he began. "The health of the environment is connected to the health of our traditional lifestyle, our traditional relationship to the land." To illustrate, Micah told me about the effects of the harbor seal population on the endangered Lake Ozette sockeye salmon population, an anecdote that had become a well-circulated talking point for tribal council members. "For over thirty years, we've been working on the recovery plan of the Lake Ozette sockeye [salmon]," Micah began. "Lake Ozette was, y'know, a really important part of our traditional territory. In the 1950s, there was a National Forest Service guy or somebody thinking that, y'know, making the Ozette River navigable would be a good thing. . . . So [he] took all of the woody debris out of the river system, and the level of the lake dropped three feet [and] exposed acres and acres of spawning ground for the salmon. . . . Logging practices [had already] decimated the watersheds that supported the river spawners. Now, the two largest limiting factors on the recovery plan are unmanaged seal predation on an endangered species and the logging practices."

"What normally held the seals in check?" I asked in response. "Bears?"

Micah had been carving a canoe paddle out of yew wood as he was telling the story, and when I asked this he stopped, looked up, and smiled.

"Us."

Stories like this naturalize the place of Makah hunters in the local environment, making their activities not just culturally and legally defensible but also ecologically necessary to the continued health of the ocean ecosystem.[11]

The Affinity's Antiquity

The perceived affinity between modern resource management and Makah traditional discourses on whaling and sealing might be explained as the product of assimilation and adaptation after more than a century of violent pressure to conform to non-Native ways of being.[12] The regulation of fishing and hunting in particular has been a contact zone between Makah leaders and the US federal government since the signing of the Treaty of Neah Bay. Federally mediated conflicts over access to marine resources raged between the state of Washington and Northwest Coast tribes throughout much of the twentieth century. The bureaucratic government-to-government relations introduced by the Indian Reorganization Act of 1934 intensified with the fisheries management reforms following the Boldt decision, when the US Ninth Circuit Court ordered federal agencies to manage the fisheries directly following the state of Washington's failure to do so fairly.[13] As a result of this intermingling and the continued presence of federal agencies near Neah Bay, many of the most influential persons on the Makah Reservation possess a vast amount of experience working within modern fisheries and forestry management systems. The Makah Fisheries Department, which is charged with regulating marine mammal harvests as well as tribal fisheries, operates within a bureaucratic framework that is patterned after—and designed to work in tandem with—state and federal fisheries management agencies. Tribal fisheries managers conduct more stock assessments each year than the state of Washington. Even Makah fishermen with little or no government experience are, by necessity, deeply acquainted with the ways and means of this system and, by extension, with federal fisheries management.

A vibrant local discourse suggests, however, that the affinity is more than just the result of assimilation or of decades-long cooperation between Makah leaders and the US federal government. Ann Renker speculates that the skills that were needed to manage seasonal food resources prior to contact with Europeans prepared Makah leaders to be effective modern resource managers. Dave Sones also suggested that the Makah concern with managing limited resources predates western conservationism: "Historically,

Figure 4.1 Makah canoe with commercial fishing fleet in background
Photo by the author

western culture didn't believe that it *could* overexploit resources, that they were so vast and so huge that there's just no way we could ever fish out the ocean or cut down all the trees, where our understanding was that it is very easy to overcapitalize resources if you're not careful."[14] Gary Ray told me that the "old folks" commonly stayed away from the areas with young halibut, for instance, which they called the "chicken banks." If fishermen came back with a catch of undersized fish, the elders would chastise them and say that they had been "robbing the cradle." Gary told me that when a group of young Makahs moved to the reservation from the Los Angeles area in the 1980s and regained the right to hunt after a period of residency, the elders noticed that they would return from the woods with half a dozen or more deer at a time, all of them fawns and nursing mothers. "So, we had an education process," Ray said, to bring the hunters back into step with the local values. "We're forced to put rules into writing now, where we never used to. We never used to have to specify anything, it was put into you, you just don't do that."

Where the Spiritual Part Begins

Historians have concluded that conservation was foreign to many Native American societies prior to contact with Europeans (Krech 2000).[15] Contrary to the myth of the noble ecological Indian, Native American societies engaged in diverse ecological practices, not all of which showed a concern for what might now be understood as best conservation practices. Remarking on the historical patterns of resource use among the Makah and their ancestors at Cape Flattery, Reid (2015) notes that while "the People of the Cape believed they were responsible for maintaining a balanced relationship with a community that included the very animals and fish that they harvested . . . we should not construe this as Makahs acting like proto-ecologists. Hunting thousands of seals and harpooning dozens of whales annually illustrate that Makahs cannot be stereotyped as Ecological Indians" (148).

My intent in this chapter is not to join the debate over whether the myth of the ecologically noble Indian has any basis in fact, but instead to point out how fisheries science provides people with a space and a register in which to claim identities based on wise ecological stewardship and lethal relationships of use, all while emphasizing ontological continuities with the ways of revered ancestors.[16]

A sense of how such talk might be born of more than political necessity became apparent to me near the end of a long conversation I had one summer afternoon with the then-director of Makah Fisheries. We had been speaking for nearly an hour around a small conference table in the main office of the tribe's fisheries department, where I had been lingering for the better part of three days waiting to speak with him. It appeared to be one of the busier times of the fishing season. An unexpectedly strong catch by a neighboring tribe had threatened to shut down one of the fisheries, throwing things into chaos and provoking a series of emergency meetings and conference calls. When the emergency finally subsided, the director sat down with me to explain in detail how the Pacific Fisheries Management Council and International Pacific Halibut Commission develop the quotas that are assigned to the Makah commercial fleet. Quoting an astounding array of figures and percentages from memory, the director spoke of halibut and salmon and whiting quotas, 80/20 splits between "restricted" and "unrestricted" fisheries, the 17.5 percent "OY" (optimum yield) of one target fish species, five-hundred-pound "trip limits," forty-eight-hour "openers," and the twenty-five-mile-per-hour winds and twelve-foot "combined seas" that were the maximum allowable inclement weather conditions in which the halibut fishery could open. At some point, our conversation turned to a prospective new tribal fishery that would target hagfish, better known to local fishermen as "slime eels." The director had recently met with a duo of Korean fish buyers who were interested in starting a hagfish fishery in the region.

"Who would have thought someone would want [to eat] a slimy eel?" the director asked rhetorically. "[But] there is a big live market for them. So the trick is getting them [to market] alive." He then told me that the "slime" that gives the eels its name is a kind of self-defense that is triggered when the eel experiences stress.

Thinking of boats bound for East Asia filled with stressed, panicked hagfish, I asked the director, somewhat clumsily, if he ever "felt bad" when he thought about hagfish being caught and shipped alive to distant markets.

"Do I ever feel bad?" he repeated.

He thought about the question for a few moments and then quickly dismissed the possibility. "No. I wouldn't think about the suffering of a slime eel." He went on to explain that he believed God put people on Earth to be stewards of natural resources, and he was not going to feel bad if there was a "resource there that could be more fully utilized." He was, however, concerned with some elements of the fish buyers' proposal. Because hagfish were an unmanaged species at the state and federal levels, Makah fishermen could conceivably fish every day of the year and catch as many of the eels as possible; this is exactly what the would-be hagfish buyers urged the director to authorize. "But that's not how we do things," the director said, and I noticed a swell of professional pride in his voice as he continued. "It never has been. It's not who we are." On the East Coast of the United States as well as off the coast of Japan, hagfish had been overfished to the point of collapse. "That's not going to happen here," he said, his voice reaching a subtle emotional crescendo. "This fishery will be *managed*."

With that sentiment, the director exhibited a concern for conservation that went beyond what was required by federal regulations. Slime eels, unprotected by any law, would not be exploited without limitation. It was, he would later tell me, where the "spiritual" part of his job really began.

The "Is" and "Ought" of Resource Co-management

"Everyone talks about ecosystem-based management now," the Director of Makah Fisheries once said to me, "but we've been saying that for years in this office. In fact, pretreaty, that's what Makahs were doing, taking an ecosystem approach and showing respect for how everything relates, as stewards."

Chad Bowechop, the chairman of the Makah Tribe's Office of Marine Affairs, offered a similar narrative on precontact resource stewardship preparing Makah leaders to be effective resource managers, but he cautioned against taking the parallels too far.

"I don't want to overromanticize Indians or tribal beliefs," he said, "because that's as big a problem as being biased and saying we don't have any rights. We were able to manage our resources at a time when resources were absolutely plentiful. It was easier to keep the balance. Were we perfect or hyperconservationists? Not at all. We were trying to survive. . . . But we did have the common sense at the turn of the century that we had to stop whaling, there weren't enough whales left." In describing his own spiritual beliefs, Chad's language is adorned with concepts drawn from fisheries management: "I believe resources are placed on this Earth by the Creator, not only to benefit man, but to benefit the rest of the world. We all have our operational place, and if you conduct yourself appropriately, there's a flow, I believe there's a rhythm on this Earth. But there are many impacts to interrupt that rhythm, or to affect it. I have strong feelings about overharvesting. You shouldn't do it."

Words like "harvest" and "resources" are easy to pick out, but "operational place" also has its origins in ecosystem-based management. "Impact" is a commonly used term in the resource management vernacular, and Chad's use of the term here calls to mind the environmental impact statements required by the National Environmental Policy Act.

There are limits to the overlap between modern fisheries management and contemporary Makah spiritual beliefs, however, and Chad warns against reducing one to the other. "We have a very sophisticated fisheries management capacity," he noted. "Does that mean strong science now is the altar we worship at? We have a lot of problems with that."

The problems begin with discrepancies between federal regulatory designations and what Makah fishermen perceive to be the reality on the ground, which is a major source of tension between the two ways of knowing the ocean. Federal scientists' reliance on systematic observation and statistical modeling means that certain species or stocks can remain on the Endangered Species List even after Makah fishermen observe signs of recovery. Accordingly, Makah officials sometimes use the term "listed" rather than "endangered" to refer to officially endangered animals that they suspect no longer deserve the designation. Federal leaders were responsive to the Makah petition to delist the Eastern North Pacific gray whale in the late 1980s, and only since the turn of the twenty-first century have these epistemological differences produced a sustained problem for Makah whaling (although similar tensions have been simmering with regard to fisheries for quite some time). "Our sacred tradition is all wrapped up in red tape," Micah McCarty said of the whale hunt as early as 1998. In 2003, McCarty wrote in the *Peninsula Daily News* that "tainted evidence" was being used to "discredit the management ability of the Makah and our federal co-managers." What might have been a temporary delay and a nuisance has become what Jo Jo McGimpsey now calls a "bogus bureaucratic obstacle."[17]

For all the apparent frustration expressed by Makah officials over bureaucratic delays in the reauthorization of whaling, it is important to observe that these complaints do not take aim at the core values of conservation science but at activists' interference with sound management practice in the service of their ulterior (read: moral, or preservationist) motives. The limitations that these bureaucratic obstacles place on a totally unrestricted Makah resource sovereignty notwithstanding, the policies that Makah leaders truly deplore are those like the Marine Mammal Protection Act that depart most dramatically from the prevailing, utilitarian approach to natural resources management employed by the US federal government. Along with the Bald and Golden Eagle Protection Act, and with the possible exception of the Wilderness Act of 1964, the MMPA is the closest the United States has come to enshrining a version of preservationist thinking in federal law, even if it ultimately failed to do so in substance as well as in its implementation (see the next chapter).[18] The MMPA, in other words, is the piece of legislation that lies closest to the values of antiwhaling activists, at least in spirit. Consider one former Makah Fisheries manager's thoughts on the MMPA and ESA:

> I think that is one thing that is really not understood in the Endangered Species Act and the Marine Mammal Protection Act, that at some point humankind needs to come back in and establish its role of utilization of these species, particularly when we've impacted other resources such as fish and the food sources those species use have been minimized, [whales] can actually start creating a negative effect on certain species because they are now the dominant population.

Notice that the manager's complaint here is that these policies are *not utilitarian enough*. For the MMPA and ESA to be effective, he argues, certain species must be opened up to further use.[19]

Makah officials are able to express this critique in management terms, including their critique of what they call "single-species management." Harbor seals play the villains in Micah's story about the endangered Lake Ozette sockeye salmon and in similar stories that are continually retold around this theme. Without Makah seal hunting, the seals threaten devastation of fish stocks. A living, swarming testament to the need for "multispecies management," the overabundance of seals shows the improvidence of policies that "ignore nature's realities" (McCarty 2003) and that fail to appreciate Makah hunting as a key and indispensable element of ecosystem health. These objections mirror debates already taking place within fisheries science, as NMFS itself publicly aspires to an ecosystem-based management approach in all its ocean fisheries (despite the fact that no one has been able to figure out how to develop the complex models of multidirectional trophic relationships needed to accomplish this).[20]

Translating Whaling

The language and logics of modern fisheries management provide an effective vehicle for translating Makah identities and tribal agendas not because these discourses are similar in every respect but because of the nature of their differences. As they impact the Makah whaling conflict, the substantive disagreements are largely epistemological rather than ethical; they do not concern whether whales ought to be killed but, rather, how to ascertain the optimum number of removals and the timing of hunts.

Scholars who are skeptical of co-management for its ostensibly neocolonial character downplay the practical importance of this moral convergence, of a shared utilitarian ethos that makes the prudence of whaling a strictly quantitative question for Makahs as well as the state.[21] There is a moment in the Makah Whaling Commission documentary on whaling that inadvertently shows the relative ease with which ideas can move between these registers—that is, between everyday language in the Makah community and the technical dialect of federal fisheries bureaucracy. Consider the following two excerpts from the film. The first is from Sadie Johnson, a Makah elder and the mother of whaling captain Wayne Johnson. The second is from Keith Johnson, president of the Makah Whaling Commission (no relation to Sadie or Wayne):

> What's all the stink about? [Laughs.] You know, I *still* don't understand why they [antiwhaling activists] don't like it. We're not hurting anybody! We only take what we need! We've been brought up that way, even if we go fishing it is for our own use, just like we'd buy in a store. We just get what we need. And that's the way we'll get our whale. Only what we could use and none of it will go to waste.

> This population, of which the International Whaling Commission has a scientific committee that does all the monitoring of this [Eastern North] Pacific gray whale population, they have all the science they need on this population in terms of numbers. It's over twenty-four, twenty-five thousand strong, today. This population can be harvested by the standards of the scientific committee of the International Whaling Commission at a number, at a tune of 484 whales *per year* can be taken out of this stock and it *will not* affect its growth.

The figure that Keith cites, 484, is almost the exact figure the IWC Scientific Committee determined to be the sustainable annual number of "takes" for the Eastern North Pacific gray whale at its meetings the same year, calculated at 90 percent of the maximum sustainable yield for the gray whale population based on the available estimates at the time.[22] Keith sees the number as an argument ender. Any further objections by antiwhaling activists would be nonsensical by the standards of sound fisheries management, given the magnitude of this number compared with that of the Makah Tribe's proposed harvest. Keith's comment not only encodes Sadie's tacit anthropocentrism ("We're not hurting anybody") into the legitimate discourse of conservation (whales are resources to be harvested) but also mirrors Sadie's sincere failure to grasp how antiwhaling activists could possibly oppose the hunt when there are plenty of gray whales and when Makah hunters will only take what they need. Notice how easily Sadie's thoughts are translated into the authorized and authoritative register of federal fisheries management.

The advantages that follow from this kind of successful translation are numerous and substantial. Scholars of indigeneity have tended to focus on the heavy symbolic costs that indigenous peoples pay when seeking legitimacy from the colonial nation-state, apparently guided by the assumption that recognition is a gratuitous imposition and it would be better if indigenous groups were simply left alone to manage their own affairs. In the Makah whaling conflict, however, Makahs have recognized that federal legitimacy offers the potential not only to legitimize whaling but to delegitimize antiwhaling, making activists' arguments nonsensical and their attempts to disrupt the hunt illicit. Micah McCarty was thus able to write an op-ed in the *Peninsula Daily News* asserting that the "emotional" claims of antiwhaling activists "ignore nature's realities" and undermine "realistic conservation necessity."[23] Janine Bowechop (2004) writes, in an edited volume on Northwest Coast ethnology, that the attempts of whaling opponents to disrupt the 1999 hunt were "an affront to US law" (418).

The advantages of this successful translation are not just theoretical or rhetorical. The US Coast Guard confiscated numerous protest vessels as they tried to interfere with Makah whalers during the 1999 and 2000 hunts. Janine later told me she was astonished that the Coast Guard, "basically a branch of the military," had been "on the side of the Indians." At the potlatch after the 1999 whale hunt, the tribe presented the Neah Bay Coast Guard station with a large carving of Thunderbird to express their gratitude for what Janine would call, in print, their "tireless enforcement" of the exclusionary zone.[24]

Despite some epistemological tensions and the fetters that federal co-management places on tribal sovereignty, the overlap between Makah ethics of human-animal interaction and the federal fisheries management model has allowed the tribe to pursue, with federal legitimacy and protection, the revival of a practice that runs counter to the stated moral beliefs of an overwhelming majority of Americans. A 1997 opinion poll found that only 14 percent of Americans supported whale hunting under any circumstances, down from around 75 percent just twenty years earlier.[25] Against the backdrop of this precipitous decline, the tribe has been able to secure approval to hunt whales from both the International Whaling Commission and (despite the current interruptions) the US federal government.

Makah leaders are continually frustrated that animal rights groups refuse to "negotiate" on this issue, a refusal that is indicative of the paradigmatic moral differences that separate the activists' ethics from that of the state. Between Makah whalers and the state, there is no such divide.

FIVE

You Just Don't Kill Whales

"Joyce never was much of anything," Margaret intones, launching into her self-described "spiel" as the weekend keeper of the Joyce Depot Museum. "Just a logging crossroads and a place where the railways went through." Today, the town of Joyce is little more than a loose straggle of businesses on Highway 112, lying near the midpoint of the Olympic Peninsula's north shore. About ninety minutes east of the Makah Reservation by car, separated by fifty-five miles of twisting highway, Joyce is also home to Margaret and her husband, Chuck, cofounders of the ad hoc antiwhaling group Peninsula Citizens for the Protection of Whales (PCPW).

Margaret is brilliant and passionate and sometimes difficult to interrupt, with a large frame, shoulder-length gray hair, and a seemingly endless supply of intellectual energy and enthusiasm. There is also something gentle, even vulnerable, about Margaret, especially when she talks about the things she cares about intensely. Her primary job at the Joyce Depot, she tells me, is to nudge people toward the Makah museum, "the number one tourist destination on the peninsula!" Still, she can barely contain her excitement as she guides a handful of visitors each Saturday through the Depot's modest collection. Beside an old wagon are various pieces of restored logging equipment, and next to them are albums filled with old black-and-white photos of loggers posing proudly next to enormous felled cedar trees. The Depot is an ambivalent homage to a time when logging companies were clearing the land of massive firs and cedars and shipping them off by rail to feed the growth of nearby cities.

I follow Margaret into the Depot's annex so that she can show me one of the products of her restless creativity: a mandala-style painting depicting the 1999 Makah whale hunt (see figure 0.1). "There is nothing fictional on here," she begins, "and I wouldn't want anyone to construe it as insulting or mocking or making light of anything."

It is almost impossible to take it all in at once: a canvas as wide as I am tall, saturated with vivid, ocean-blue oil paint and dotted with the likenesses of dozens of people and objects. At the center is the nose of a single gray whale, poking straight up out of the water in what whale biologists call a "spy-hop." On both sides of the whale are mirrored images of the Makah whaling canoe, then the motorized tribal chase boats, federal observers on their government boats, the Coast Guard's rigid inflatable zodiacs, and a bevy of small motorized protest boats. The Sea Shepherd Conservation Society's iconic black cutter circles them all. On a green patch of land at the far edge of the frame is a protest roadblock set up by PCPW.

"These are all real people," Margaret continues. "Every sign is an actual sign that was carried, and every person is an actual person."

I had met—or would later meet—many of them. There's Keith Johnson, president of the Makah Whaling Commission, and Pat Gearin, the federal biologist whom many activists despise for his open support of the Makah hunt. "That's Paul [Watson, founder of the Sea Shepherd Conservation Society]," Margaret says, pointing to a white-haired figure. "And there's Wayne [Johnson]. Oh, and there's Will!" Margaret cries gleefully, indicating a tiny watercolor avatar of the placid Seattle-based activist and author of a soon-to-be-published manifesto on green veganism. Two boats over from Will is Jonathan Paul, a young EarthFirst! activist who led the charge to interfere directly with the 1999 and 2000 Makah hunts, and Tami Drake, a paralegal who dropped everything and moved to Washington to help the activists after she heard about the Makah whale hunt on TV. She and Jonathan met during the protest and were married soon after. "They're just the sweetest people on earth," Margaret says of the activists, suddenly sounding forlorn. "We really bonded with a lot of these characters."

Margaret's adoration for the other activists, especially the young animal rights activists who flocked to Neah Bay in the 1990s to protest the Makah hunt—whose motivations she repeatedly describes as "pure . . . much purer than us"—belies her own importance in the story. The Makah whaling conflict has touched distant shores and has drawn the attention of famous and powerful people. Yet none have been more influential than Margaret and Chuck, two "poverty-by-choice fishermen hippies" who rarely leave their home in the Washington woods. "We were probably the biggest thorn in [the Makah's] side," Chuck once told me. "No one else in the world knows this issue like Margaret. And it's kind of amazing when you think about it, how we met with governors, senators. . . . Very few people know the extent of the role we've played."

Accepted wisdom has it that antiwhaling activists tend to be urban elites whose primary interaction with nature comes in the form of wilderness-based recreation. Writing in the early heyday of the environmental movement, Roger Caras (1971) claimed that few among the antihunting lobby ever go "afield" and that "for the individual in this category wildlife is a remote concept" (89). William Cronon (1996) wonders why environmental debates so often "pit urban recreationists against rural people who actually earn their living from the land" (85).[1] These are the generous descriptions. Most previous accounts of the Makah whaling conflict have either ignored the activists or portrayed them all as racists.[2] Others attributed the outrage incited by Makah whaling to the use of modern implements in the hunt, an explanation that mistakes (misguided) rhetoric for substance (e.g., Raibmon 2005, 206).

Chuck and Margaret do not fit these profiles. Chuck is a former commercial fisherman whose livelihood brought him into regular, direct contact with fish, whales, and other marine wildlife. With a penchant for denim overalls and an enormous gray beard that seems ever poised to engulf his upper body, Chuck more closely resembles a stereotypical logger than an environmentalist. He now works delivering the *Peninsula Daily News* on a rural route. Margaret affectionately calls him "the paperboy." Margaret helps support the family by selling custom, handmade ceramic tiles. Together, they fit no one's image of disconnected urban elites.

The Sea Shepherd Conservation Society received the largest share of the media attention leading up to the first Makah whale hunt after they sailed into the Makah harbor in a repurposed Coast Guard vessel in the summer of 1996. Almost immediately, they became the face of outside resistance in the minds of Makah tribal members.[3] Led by Paul Watson, a charter member of Greenpeace, Sea Shepherd later became famous for its annual "campaigns" against Japanese whaling in the Antarctic whale sanctuary, which since 2008 have been the subject of Animal Planet's reality television series *Whale Wars*. Sea Shepherd was accompanied by smaller animal rights organizations from around the region, and others soon joined the fray: kayakers from California, whale enthusiasts from Seattle, even a billionaire who offered to buy off the Makah. A fleet of Canadian whale-watching vessels decided to use their fleet to intervene in the hunt. Some of these groups remained in the area for months, camping out in cheap motels and living mostly on donations. Their ranks swelled on weekends during what one activist affectionately called

Figure 5.1 Portrait of Chuck and Margaret Owens that appeared alongside "Opponents of Makah Whale Hunting Remain Steadfast," *Peninsula Daily News*, May 17, 2009
Photo by Chris Tucker; reproduced courtesy of the photographer

"whale wacko Saturdays." She was referring to new-age spiritualists who ate cheeseburgers and felt whales were "the brains of the universe," a combination that many of the animal rights activists found ridiculous.

One by one, these groups left. By the time a federal court suspended Makah whaling in 2000, only a few organizations remained. Some have since moved on to other battles, while others have disbanded in the wake of the expanding criminalization of animal rights activism in the United States.[4] In 2002, Sea Shepherd declared that the Makah whaling conflict was a "local issue."[5] Today, only Chuck and Margaret remain, at the vanguard of antiwhaling activism in the Makah whaling conflict as they have been since 1998.

The motives and early tactics of Chuck and Margaret's activist group, along with the rise and eventual decline of what I call "interventionist activism" in the Makah whaling conflict, are the subjects of this chapter. While the next chapter tells a story of subtle disciplinary coercion by exploring the activists' attempts to translate their objectives into the language and logics of fisheries management, the story begins here with activists' emergence and eventual defeat at the hands of a more conventional form of state power.

It Was in the Air

Global attitudes toward whales and dolphins underwent a remarkable transformation between the 1960s and 1970s. For much of the first half of the twentieth century, whales remained objects of a rapacious albeit declining global industry, a situation that provoked little outrage on either side of the Atlantic and gave little indication of whales' future status as global icons of the environmental movement.[6] The conservation of the large rorqual species that inhabited the Southern Ocean (blue whales, fin whales, and humpbacks) had been discussed as early as the 1920s, but early conservation efforts were largely ineffective, as industrial whalers from Japan as well as Norway and other northern European nations successfully fought off any meaningful international catch restrictions. Driven by nationalist pride and visions of economic autonomy or empire—what Charlotte Epstein calls "whaling nationalisms"—the Dutch and Japanese whaling industries remained afloat even after their commercial viability had become questionable.[7] In swarms of catcher vessels around so-called floating factories, whalers from these and other European nations continued to kill around 25,000 fin whales each year, primarily around Antarctica, until as late as 1962. The International Whaling Commission, having formed in 1946 with the aim of brokering international agreements that might limit catches to sustainable levels, was burdened by infighting for much of the first two decades of its existence, as whaling nations weaponized scientific uncertainty in order to delay catch restrictions.[8] By the time the IWC introduced national quotas, "so depleted were the whale stocks that these could not even be filled" (Epstein 2008, 40).

Things might have continued in this manner until the commercial extinction of all of the large whale species if not for a shift in broader public attitudes toward whaling.[9] A key factor in this transformation was the advent of underwater photographic and acoustic technology. Underwater images of whales were not widely available prior the 1960s, before which Leviathan had never fairly floated himself for his portrait. Available illustrations were based on specimens that either washed up on the beach or lay upon the flensing decks of whaling vessels, and they depicted whales as lumpy and amorphous. New underwater footage revealed whales to be far more fluid and graceful than these renderings had suggested.[10]

It was also during this time that cetaceans first appeared as actors in film and television, beginning with the 1963 film *Flipper* and its television adaptation. The first SeaWorld amusement park opened its doors in 1964, with its most popular show built around the intelligence and charisma of Shamu, the first killer whale to survive more than thirteen months in captivity. Whales' public profile was further boosted by Roger Payne's discovery and recording of humpback whale song in 1967. Payne's 1970 album *Songs of the Humpback Whale* remains the best-selling natural history record of all time.

Of the underlying factors that led to the global transformation in attitudes toward whales, perhaps none were more important than the captive research programs funded by the US military and its Cold War–era programs in acoustic bioscience. Historian Graham Burnett (2012) particularly credits the controversial work of American neuroscientist John C. Lilly, whose experiments on cetacean intelligence began as part of his work for the US Navy (622–45). Extrapolating from his findings on bottlenose dolphins, Lilly claimed that large cetaceans like sperm whales possessed mental and linguistic abilities that could rival or even exceed those of human beings. While many of Lilly's claims have since been discredited (530), his work has had an enduring impact. A 1971 US Senate joint resolution that called for a global moratorium on whaling cited Lilly's work a dozen times (625). It was Lilly's lab that helped refine the bioacoustics technology and methodology that enabled humans to hear the varied songs and communicative sounds of whale species that had been long believed to be mute (635–37).

Whales were well established in the American popular imagination when the environmental movement reached the early peak of its political visibility, beginning with the first Earth Day in 1970 and the first United Nations Earth Summit in Stockholm in 1972. Whales featured in two of the signature legislative actions of the so-called environmental decade of US policy reform: the Endangered Species Act (which listed all large cetaceans as endangered until further data could be obtained) and the Marine Mammal Protection Act. Perhaps more significantly, whales became *the* symbol of the global environmental movement, with narratives of their plight serving as a metonym for a broader environmental crisis. In 1975, a Canadian environmental group inspired by Lilly's writings attempted to stop a Soviet sperm whale hunt off the coast of California.[11] Riding in rigid inflatable Zodiac boats, the activists attempted to interfere with the hunt by placing themselves between the sperm whales and the Russian catcher vessels. The whalers fired a harpoon that traveled mere feet from the activists' heads, and footage of the incident was broadcast on American evening news, sparking outrage and marking a watershed moment in the environmental movement.

You Just Don't Kill Whales

Whatever the final tally of its causes, the transformation in attitudes toward whales and dolphins in most of North America and Europe was swift and significant. As late as 1960, there was very little popular interest in these animals in the United States, but by 1975, whales had become global icons of the environmental movement. For Margaret Owens, the connection between these shifting attitudes and her own later activism is apparent. "Just being of that generation," she said of those who were teenagers in the 1960s, "it was in the air." Margaret describes her opposition to whaling as "gut knowing/feeling" coming from a confluence of factors, a "gestalt" of emotions that she cannot easily explain: "Everyone, everywhere . . . if you were out on the boat and a whale surfaced and blew you'd just go 'Ah!' There's just something about [seeing whales] that defies explanation, to be close to something that big, that living, that harmless . . . or just to see it. It's like climbing the peaks to see that one biggest peak, and the way it makes you feel. It's the *aesthetics*." She added later, "What other animal can captivate people with its very breath?"

The same breath, the same enormous, salty exhalation that is for some Makah fisherman a stomach-turning experience is, for Margaret and other activists, magnificent—an independent reason for protecting whales and drawing lines between them and other species.[12]

"There's something there that's hard to explain," Chuck added. "Something *majestic*."

Chuck shares the sentiment but arrived at it through different means. His strict, working-class upbringing insulated him from environmental activism in his youth. He likes to tell Makahs that he has probably eaten more whale meat in his life than they have, thanks to his adoptive brother, an Alaskan Native who brought whale meat into the family home throughout Chuck's childhood. As a young fisherman, Chuck remembers being awestruck at the sight of whales. "It was the most exciting thing that could happen on a fishing trip," he recalls. He almost knocked himself out on the low ceiling of the ship's cabin on two different occasions after a glimpse of a whale spout caused him to jump quickly out of his chair.

On his very first voyage as a deckhand on a fishing boat, Chuck stopped the skipper from shooting two pilot whales. The skipper had claimed that

Figure 5.2 Chuck Owens, ca. 1980
Photo by Margaret Owens

the whales were scaring all the fish away. "I knew nothing about whales [at the time]," Chuck said. "But I went up to him and said, 'As long as I'm on your boat, you're not shooting a whale.' That was pretty much the end of my fishing career for him." Once he had his own boat, Chuck wouldn't allow his crewmen to shoot any "living creatures."[13]

Chuck sometimes struggles to reconcile his activism with his eating choices and the fact that, despite his misgivings about killing animals, he is not a vegetarian. "I had to draw the line with [whales]," Chuck explained to me. He listed a handful of reasons, including whales' intelligence, the size of their brains, their use of "language," and their ability to "know *their* world just as we know ours."[14]

Once Chuck and Margaret were married, whale sightings and whale imagery continued to play a salient role in their lives. Their first joint financial act was to trade an old car for a huge block of soapstone, out of which Margaret carved a sculpture of a mother and baby whale. This first sculpture seemed to combine the traits of several species of whales.[15] Soon, Margaret began sculpting gray whales out of clay. She recalls making dozens of sculptures of various sizes at a time when the public knew little about the species. The images that were available at the time depicted them as lumpy and bulbous, and Margaret's early art reproduced the error. In 1984, the local newspaper printed a picture of one of Margaret's sculptures, a clay gray whale she now describes as "turd-like."

To make better sculptures, Margaret was compelled to learn more about gray whales, and doing so made her feel more and more connected to them. "Living here on this little tip of the peninsula," Margaret said, "what we have are gray whales. We have them close to shore, and we have them where we can see them, and where our kids can see them, and when we're at the beach we can see them. Familiarity breeds caring." Chuck fondly recalls a memory of their youngest child pointing out from the beach and exclaiming, "They're sprouting!" (spouting). When Chuck would return home after months at sea, the most exciting thing he could tell his children about was seeing whales. "We just had a history with the whales," Chuck said, as if he were talking about a dear friend of the family. "They were just a part of our lives." The sum of these feelings and experiences and the primary impetus for Chuck and Margaret's opposition to Makah whaling is reflected in a moral judgment that they were shocked to discover not everyone on the US mainland shares: "You just don't kill whales."

Shooting from the Hip

When Margaret and Chuck first heard of Makah plans for a whale hunt, they were unconcerned. There was simply "no way," they thought, that the government would allow the tribe to kill *whales*. In 1998, the couple attended a talk by Will Anderson, the director of the Washington-based Progressive Animal Welfare Society, at a local community college. Anderson described visiting the gray whale birthing lagoons in Baja California in the 1970s, before it became the whale-watching destination it is today. He would camp there for weeks, so close to the lagoons that he could hear the whales breathing at night. Will came to Neah Bay when he heard that the International Whaling Commission had approved the Makah Tribe's plans for a whale hunt. Moved by Will's passion and shocked by the suggestion that someone would hunt whales so near to their home, Margaret and Chuck quickly organized a rally in nearby Port Angeles to "support the whales." The rally formed the basis of a small protest group that would later become PCPW. The group was comprised mostly of local "animal lovers," many of them elderly women.

The decision to form a protest group was not without difficulty for Chuck and Margaret. Both had long been supporters of Native American causes in the area. Both would later help fight alongside the Elwha Tribe in 2004 against plans for a state-backed construction project that would have destroyed an ancient tribal burial site.[16] They feared being associated with "racist" agendas, including those of protesters who opposed Makah whaling out of bitterness over Indian gaming laws or tribal fishing rights. "There's wrong reasons to oppose [whaling]," Margaret once told me, "and there's right reasons to oppose."

"That's weighed heavy on us," Chuck added. "We took on a tribe that in most cases we would support. . . . But somebody had to speak for the whales."

Margaret and Chuck had little experience as environmental activists prior to the formation of PCPW. Their first formal action was to support a letter-writing campaign to the Makah Tribe, a tactic Margaret would later describe as naive. "We really believed," she said, sounding at once wounded and self-critical, "that if enough people just asked [the tribe] to please not do it, that they would consider not doing it." When it became clear that the tribe would not acquiesce, Chuck drove to nearby Port Angeles (the largest town in the area, with a population of around nineteen thousand) to try to convince business owners to post messages opposing the hunt on their reader boards. "We were just shooting from the hip at the very beginning," he told me. "We didn't know what to do."

No one remembers how the group first decided to hold a protest near Neah Bay, but soon protest roadblocks at the Makah Reservation border became PCPW's trademark public action. Every Saturday for several months, the group would meet at Sekiu, a small town nineteen miles east of Neah Bay, and drive toward the Makah Reservation in a long caravan until Chuck's car conveniently "broke down" near the border. Blocking traffic, the group would hold up signs and chant for twenty to thirty minutes until the county sheriff told them to break things up. Margaret held a white rectangular sign at the protests that read "HONK FOR WHALES" in bold capital letters. The sign's message elicited a range of responses from motorists. In a preview of the kind of rational data tracking that would become one of PCPW's hallmark tactics, Margaret made tally marks on the back of her sign, sorting the responses of each and every passerby into categories. The master category was "honk" versus "non-honk"; the other categories included "stink-eyes" and "middle fingers." She wrote down everything anyone yelled at her, no matter how "colorful."

From the beginning, Chuck claims, PCPW's tactics were different from

Figure 5.3 Roadside protest, 2002

other groups'. "We live here," Margaret emphasized to me on many occasions in our interviews. "Other groups, they get to go home. But we intend to live here our whole lives." For this and other reasons, they aimed for transparency in their public demonstrations. "We kept everything aboveboard," Chuck said. "We didn't want no surprises." Before each roadside protest, Chuck would inform the Clallam County Sheriff of their plans, including the time and rough location of the demonstration. The sheriff would then notify Makah tribal police, who would set up the roadblock. Still, the roadside protests were dangerous. It was already the fall of 1998, and tensions were high. The whaling crew had reportedly been training for months, and the tribe had received international and domestic approval for a hunt. Adding to the volatility, Chuck and Margaret's group was soon joined by activists from Seattle and elsewhere, about whom PCPW knew very little. A protester from Canada attended the demonstrations regularly and held up sign that read "Subsistence My A$$!"—a sign that Chuck and Margaret approved. But another activist held a sign that read "Boycott Casinos," which they both felt crossed a line. Chuck made a point of getting on his loudspeaker and announcing that this latter individual was not a part of their group.

The roadside protests drew the interest of reporters who were already stationed near Neah Bay in anticipation of the hunt. Some tribal members would go down to the barricades as well, and occasionally, shouting matches ensued. Robert Sullivan describes an incident at one of the demonstrations where a woman repeatedly screamed at Wayne Johnson that he was "evil as hell." Margaret recalls a Makah man screaming at the activist in that same moment, "I'm gonna put your head on a stick in front of my house!" It was not until later, Margaret said, when she was reading about Northwest tribal history, that she realized it was a "culturally correct threat."[17]

Direct Intervention

Around the same time PCPW was holding its land-based protests and while reporters in Neah Bay were waiting for a whale hunt to begin, a flotilla of antiwhaling activists arrived by sea with the intention of disrupting the Makah hunt.

Paul Watson anchored the *Sirenian*, a repurposed Coast Guard cutter that had been repainted jet black, outside the Neah Bay harbor in late summer. Tribal members referred to it as "the black boat," and those who were children at the time remain utterly convinced that it had functioning deck-mounted cannons that were pointed at the town the entire time. Sea Shepherd blasted whale sounds through speakers on the deck of the *Sirenian* in protest of the coming hunt, and Wayne Johnson later told me that the activists would yell out menacing messages over a loudspeaker, taunting a tribal official about a past drug conviction and asking Johnson if he wanted some marijuana. A second Sea Shepherd vessel carried a thirty-six-foot submersible craft from which they intended to emit killer whale calls to frighten away gray whales, but a federal court ruled the strategy impermissible under the Marine Mammal Protection Act.[18]

In a ragtag fleet of smaller vessels, a group of young animal rights protesters prepared their own form of activism. Loosely affiliated with the Animal Liberation Front and EarthFirst! movements, theirs was a model of direct intervention. One of the activists was Jonathan Paul, a self-described "eco-animal activist" and "deep ecologist" who would later serve thirty-eight months in prison for his role in the destruction of a commercial horse slaughterhouse. A vegan since the age of seventeen, Jonathan believed it was his destiny to be an activist after viewing videos of vivisections while in high school. He joined the antivivisection movement in its early stages and was active throughout the 1980s and 1990s. Jonathan helped found the first "hunt sabotage" campaigns in the United States, a tactic that focuses on directly disrupting big game and trophy hunts.

Jonathan first heard of Makah whaling from Paul Watson, whom he had known since the early 1990s, and he decided he would try to intervene. His organization, Sea Defense Alliance, or SEDNA (later renamed Ocean

Defense International), had experience protesting shark fishing in the ocean waters off California.

"For me it was real simple," he said. "We *just* stopped commercial whaling. I was like, there's no way that's gonna happen here. And then I started hearing more about what was really behind [Makah whaling], that it was really the Japanese trying to bolster their own whaling, and the Makah were seeing dollar signs and wanted to build a whaling station." Jonathan claimed to be sympathetic to the Makah on the issue of treaty rights, but he was more interested, he said, in "the culture of the whales." He was surprised when another EarthFirst! activist approached him at an animal rights conference and exclaimed of Makah whaling, "But it's their culture!"

"Why is suddenly the human culture always more important than the [way of life] of other species?" Jonathan asked. "Why do we always have to be so human-centered about everything?"

For several weeks in the fall of 1998 and for nearly three months in the spring of 1999, Jonathan and his associates stationed themselves in the nearby town of Sekiu, about thirty minutes by boat from Neah Bay, and sailed each morning to the hunting grounds. The waters around Neah Bay can be especially rough and dangerous for small boats in the fall and winter, and Jonathan's crew was inexperienced. To make matters worse, he was piloting a small speedboat and its single engine kept breaking down. "It was almost comical," he said. The Makah government suspended the fall whale hunt without success (the migration of whales came late, and the weather became unfavorable), and Jonathan was able to raise enough money over the winter to invest in a swifter forty-foot jet boat and a nineteen-foot rigid inflatable Zodiac.[19]

Jonathan describes the protests in 1999 and 2000 as a frenzy of quick action and high-stakes maneuvering. Every morning at three o'clock he would awaken his crew and cruise toward Neah Bay in total darkness. The predawn trip was dangerous because of the many submerged rocks and other hazards near the shoreline. "You can't see anything," Jonathan recalls. "You just *hope* you don't hit anything." Once they arrived at the hunting grounds they invariably found themselves at a disadvantage compared to the Makah chase vessels, whose pilots knew the water intimately. Jonathan remembers one episode in which he suddenly realized one of the pilots was trying to lead him onto a submerged sandbar. "It was a good strategy," he said. "I would have done the same thing."

As the activists in small boats tried to intervene by placing themselves between the whales and the Makah canoe, Sea Shepherd circled at a distance in their larger vessel. Chuck and Margaret joined the young activists on the

Figure 5.4 High-stakes maneuvering, May 15, 1999—Vessels from left to right: the Makah whaling canoe, a protest boat, a NOAA Fisheries boat, Sea Shepherd's *Sirenian*
Photo by Margaret Owens

water during the 1998 and 1999 hunts. Chuck's experience as a fishing boat captain made him a valuable pilot. The tactic found its limitations, however, when the Coast Guard established its moving exclusionary zone around the Makah whaling canoe. Any protest vessel that came within 500 feet of the whalers was swiftly confiscated. Jonathan's boat was seized by the Coast Guard on May 15, 1999, two days before the Makah successfully killed a whale. The charge warranting the seizure, ironically, was that Jonathan had violated the MMPA by getting too close to the whales. By the morning of May 17, most of the activist boats had been confiscated.

Jonathan's organization resumed these tactics during the 2000 family hunts, this time using lighter, faster personal watercraft that were less costly to lose. One of Jonathan's crew members was seriously injured, however, when her personal watercraft was run over by a Coast Guard boat. The Makah whaling support crew was required to broadcast an announcement every thirty minutes on a specific channel while they were actively hunting in order for the moving exclusionary zone to remain in effect. Whenever the whalers missed an announcement, Jonathan and Tami would send someone in on a Jet Ski to get between the whale and the canoe. In this instance, the Coast Guard did not recognize the purported loophole, and they collided with the young woman's personal watercraft. They then pulled her from the water by her fractured arm and handcuffed her, according to Jonathan and Tami, and when they tried to approach to render aid, the Coast Guard told them not to come any closer or the same thing would be done to them.

Legitimate Violence

If the antiwhaling activists involved in the Makah whaling conflict shared one thing besides their opposition to whaling, it is that few felt compelled at the outset to articulate a justification for their opposition to the Makah hunt. In fact, for Margaret Owens, one of the first major surprises was that the intuitive feeling that led her to oppose whaling was not universally shared, and that even if it was, simply bringing attention to the issue was not sufficient to stop the Makah Tribe's plans. PCPW's roadside protests, while an effective way to raise local awareness about the issue and "blow off steam," were also somewhat redundant: it did not seem to matter how many people were opposed to whaling as long as the tribal government wanted to move forward with the hunt and the National Marine Fisheries Service supported it. An overwhelming majority of Americans already opposed whaling, yet the Makah had been given permission to hunt off the coast of one of the most politically liberal (and whale-loving) states in the country and had enjoyed the protection of the Coast Guard in doing so.

The continued evolution of PCPW's tactics, which I describe at length in the next chapter, reflects their realization of these facts and the slow retreat of the group's optimism regarding the status of cetaceans before the law. Interventionist activists like Jonathan Paul were already convinced that no amount of protest or negotiation could bring a halt to the tribe's plans, which is why his group sought to interfere directly by placing human bodies between the whale and the harpoon. While this tactic might have contributed to the tribal whaling crew's inability to harpoon a whale in the first few days of the 1999 hunt, it was rendered ineffective when most of the activist boats were confiscated by the US Coast Guard.

The failure of interventionist activism highlights an important part of the story of state power in the Makah whaling conflict. Scholars have shown how technocratic authority works to co-opt dissidents subtly, deploying expert knowledge to create high barriers to entry that preempt political discussion (see, e.g., Wynne 2010). This kind of power certainly plays a substantial role in the Makah whaling conflict, which the next chapter will explore in detail. But the activists' boats were seized not by ideas or technocratic barriers but by sailors in uniforms with boats and guns who were authorized by the state to use violence if necessary. It is perhaps incidental that when Jonathan Paul and Tami Drake approached their injured fellow activist as she was taken into Coast Guard custody, an officer on the Coast Guard boat allegedly warned that if they continued to approach, the same would be done to them; but it is inarguable that the activists had encountered, in that moment, a rather conventional form of state power: one underwritten by the threat, and perhaps the spectacle, of legitimate violence.

SIX

The Science Has Ruled

Shortly after we met for the first time, as we walked along La Push Beach during the Quileute Tribe's annual Whale Welcoming festival in the summer of 2011, Margaret Owens told me of an exciting new development in the Makah whaling conflict. She had just been notified that the results of a long-awaited study on gray whale population genetics were soon to be released. The study, conducted by multiple agencies and interest groups including the US federal government and the Makah Tribe, had found that a tiny subset of Eastern North Pacific gray whales—two hundred or so out of the estimated twenty thousand currently roaming the Pacific—could be distinguished from the rest of the population by small but statistically significant differences in the haplotype frequencies of their mitochondrial DNA.[1] This putative subpopulation—which Margaret and other local antiwhaling activists call resident whales, but which biologists and administrators for the Makah Tribe, US federal government, and International Whaling Commission call the Pacific Coast Feeding Group (PCFG)[2]—spend the summer months feeding near the Washington and Oregon coasts instead of migrating with their conspecifics to the Bering and Chukchi Seas.[3] Scientists suspected that gray whale mothers "teach" this foraging behavior to their calves and that after a few generations this "learning" would be reflected in their mitochondrial DNA. This is what the new study seemed to confirm.[4]

"The science has ruled," she told me. "It's all moot now, [all] my reams of paperwork and documentation [and] so many aspects we've attacked. I think it's all gonna, luckily, be eclipsed by the science that's gonna save these whales. And as far as the migratory whales, it remains to be seen whether the whole population can be broken down into subpopulations based on the feeding areas that the mothers brings the calves [to]."

The publication of the study was the most significant development in the Makah whaling conflict in nearly five years. Makah officials were also fervently discussing its implications. The tribe did not believe the findings spelled the end of their hopes for going whaling again, but the results did present a challenge that could set back their efforts by ten years or more. The National Marine Fisheries Service had already announced plans for a new environmental impact statement (EIS) in light of the findings, discarding years of work on a previous draft that was still in progress.

"It's not gutter science," Micah McCarty conceded shortly after the announcement of the results, defining the study in contrast to his favored term for what he considered to be biased or interested research.[5] Indeed, the tribe's own marine mammal biologist was credited as one of the authors on the study, which replicated the findings of a Canadian biologist whose research tribal leaders viewed with greater suspicion.

The issue was not that Makah whalers intended to target resident/PCFG whales, although whether they intend to do so remains a point of contention between whalers and activists today. The problem, rather, was that in the course of pursuing the more abundant migratory gray whales, Makah whalers could not guarantee that they could avoid striking or killing the smaller and more vulnerable population of resident/PCFG whales, which come nearer to shore, linger much longer, and are for those reasons easier to hunt.

The concern over resident whales mirrors an approach to fisheries management common in the waters off the West Coast of the United States informally known as "weak-stock management." The goal of weak-stock management is to adjust harvest strategies to protect the most vulnerable "stocks" while achieving what is called an "optimum sustainable yield" of the target species. Vulnerability is determined by "stock assessments" that are informed by the work of research and conservation biologists. Each user group receives a quota not just for the species they target but for all species they will inadvertently catch (the latter is known as "bycatch"). Once any one of the quotas is reached—and usually it is one of the weaker stocks that is reached first—the entire fishery for the target species is closed for the year for that user group. For Makah whaling, the new study effectively set the quota for resident/PCFG whales—in this case the weaker stock—at zero until a "potential biological removal" could be calculated, which therefore shut down all harvest of gray whales for the time being, including the more abundant stock of migratory whales.[6]

Margaret's enthusiastic response to the study shows how deeply the activists had been drawn into the discourse of federal fisheries management. By the time I arrived at my field sites in 2010, activists were focusing almost all of their efforts on regulatory science, hoping to turn an obligatory passage point into a strategic advantage. In chapter 4, I argued that a moral affinity enabled Makah officials to effectively translate their objectives into forms acceptable to NMFS without compromising their political goals or subordinating their moral understanding of whales and whaling. For activists, the equivalent step was always going to involve a more dramatic transformation. The foundational assumptions of stock-based management—in short, that whale populations represent exploitable "stocks"—are about as far as one can imagine from the stated beliefs of antiwhaling activists, many of whom entered the Makah whaling conflict in the 1990s with the goal of "speaking for the whales." This chapter examines the activists' attempts to translate their goals into the language and logics of federal fisheries management, examining the ways in which the activists' strategic adoption of the state's narrow vision of cetaceans has changed the way they advocate for whales.

To understand how far the activists needed to go in pursuit of this aim, we first need to examine the moral economy of the agency that controls the conversation.

The Moral Economy of NMFS

Since the 1970s, the National Marine Fisheries Service has been the sole agency charged with implementing three major pieces of domestic legislation as they pertain to whales and whaling, each of which emerged from the so-called environmental decade of US policy reform. The Magnuson-Stevens Fishery Conservation and Management Act, the primary law governing commercial fisheries in US waters, places the regulation and monitoring of fish, whales, and other "living marine resources" under the agency's purview. The Marine Mammal Protection Act assigns NMFS sole authority to conserve and manage whales, dolphins, seals, and sea lions, as well as to oversee the procedure for granting aboriginal exemptions to its moratorium on the killing of marine mammals; and the Endangered Species Act charges NMFS with overseeing the "recovery" of fisheries and whale stocks. NMFS is also charged with implementing the regulations of the International Whaling Commission through the Whaling Conventions Act. When Makah officials initially approached the federal government about delisting the gray whale and, eventually, about reinstating a whale hunt, it was NMFS that held the authority to grant the requests.

By referring to the "moral economy" of NMFS, I want to draw attention not to the formal ethics of the institution (cf. Merton 1938) or to the personal moralities of its human constituents (cf. Sharp 2013), but to the tacit moral consequences that flow from the agency's particular framing of the natural world. As a technocratic institution, NMFS and its agents may understand their work to be value-neutral, but as scholars have routinely shown, technocratic institutions begin with normative presuppositions that are not wholly attributable to instrumental or bureaucratic rationality (Daston 1995; Jasanoff 2012, 168). I am not suggesting that these presuppositions are conscious or deliberate; rather, the moral economy of NMFS emerges as a *consequence* of the limitations the agency places on the biographical possibilities of gray whales, which in turn limits the kinds of arguments the state will accept and act upon.

For NMFS, whales exist not as individual beings but as natural resources and fungible elements of statistical models—in short, as *stocks*. The agency's criteria for determining what counts as a stock under its implementation of the MMPA are fluid and changing and were the subject of numerous workshops and working groups convened in the early 2000s, at least some of which were occasioned by concerns about the resident/PCFG subpopulation of the Eastern North Pacific gray whale stock. What is not in question—what is never really subject to question—is that the *stock* is the proper and smallest unit of conservation. "In the world of marine conservation," anthropologist Jennifer Telesca (2017) observes, "there is no keyword more important than fish *stock*. Commonplace, taken-for-granted, repeated without end, this single term expresses the granular work of unbounded human power over life under extractive capitalism" (144). In its strategic plan, NMFS identifies a healthy, sustainable fishery as "one in which the rate of fishing mortality does not jeopardize the capacity of the stock to produce the maximum sustainable yield on a continuing basis" (NMFS 2014). The Magnuson-Stevens Act, the agency's primary charter, provides the agency with the principal objective of insuring "optimum yields on a continuing basis" and maximizing the "full potential of the Nation's fishery resources" (secs. 104–297).

NMFS draws its legitimacy from the legislation that authorizes it as well as the technical expertise it draws upon. What may be surprising, given the prominence of large cetaceans in the narrative and imagery of the US environmental movement since the 1970s, is that all of the sources of NMFS whale policy converge in their treatment of whales as exploitable resources. Of these, the most surprising may be the Marine Mammal Protection Act, since it has generally been viewed as a victory for whale preservationists—that is, for those who oppose the killing of whales and dolphins under any circumstances. The MMPA placed an indefinite moratorium on the "taking" of marine mammals in US waters, and as such it is one of the few major pieces of US conservation law to have what might be called a preservationist spirit. Yet as James Scarff (1980) points out, the text of the MMPA "in fact rejects the preservationists' essentially ethical arguments. . . . Once scientists have determined that a population of a species is at or above its 'optimum sustainable population level,' the Secretary of Commerce can permit the taking of animals from this population *for virtually any use* so long as the population is not reduced below the OSP [optimum sustainable population] level" (262–63; emphasis in original). The legislative history of the MMPA provides a succinct demonstration of the moral economy of stock-based fisheries management in action. Originally promoted by animal rights organizations, the language of the bill was ultimately made to agree with the existing natural resource orientation of federal bureaucracy. Perhaps equally important is that Congress assigned NMFS sole authority to implement the MMPA as it pertains to whales, dolphins, seals, and sea lions, including the exemption process for aboriginal whaling and sealing. Unsurprisingly, the agency implements these key provisions using a stock-based management approach.

Similarly, the International Whaling Commission is perhaps best known to the American public for its global moratorium on commercial whaling, ratified in 1982 with the heavy backing of the US delegation, but the history and structure of the IWC reveal an organization that is committed to the regulation—not the abolition—of whaling.[7] The IWC's quota system was modeled explicitly on fisheries management protocols (although not exclusively; they also took inspiration from extractive industries like mining and oil/gas exploration).[8] Until 1978, the IWC "manage[d] every stock of every species of whale on the assumption that maximum sustainable harvesting [was] the social optimal policy" (Scarff 1980, 259). Today's meetings of the IWC's scientific committee are devoted as much (or more) to the development of efficient killing methods as to whale preservation, a darkly humorous indication of which can be seen in the organization's decision to place all questions of animal welfare under the purview of a committee called the Working Group on Whale Killing Methods and Associated Welfare Issues. Nothing, perhaps, betrays this utilitarian ethos better than the commission's former standard unit for apportioning whaling quotas: the garishly abstract "blue whale unit," based on the average amount of oil derived from one blue whale (two fin whales, for instance, or two and a half humpback whales being equal to one blue whale unit).[9] Today, several of the IWC's member nations continue to whale commercially in objection to its moratorium. Until Japan's recent withdrawal from the commission, Japan's Institute for

Cetacean Research hunted minke whales in the Southern Ocean for commercial sale for several decades under the pretense of its scientific whaling program. The IWC still traffics exclusively in numerical quotas and population estimates when it considers the acceptability of scientific or aboriginal subsistence hunts.

The scientific wings of NMFS and the IWC are the institutional inheritors of what Graham Burnett (2012) playfully calls "hip-booted" cetology for its close relationship, bordering on parasitic dependence, with the early twentieth-century whaling industry (the eponymous hip boots were what was needed to wade through whale viscera on the flensing decks of whaling vessels in the Southern Ocean).[10] Those close relationships with industry no longer exist for member nations like the United States whose whaling industries have vanished, and the constant need for fresh (dead) specimens— once a hallmark of whale conservation—remains only in vestigial form, visible today in the work of conservation biologists who flock around bowhead whale carcasses on the Alaska North Slope during indigenous hunts as well as in the "scientific" research program of the Japanese Institute for Cetacean Research. Yet stocks remain the coin of the realm, and in the moral economy of NMFS (and the IWC), whales remain exploitable marine resources.

NMFS is not just the passive recipient of these policies. The agency's exclusive focus on stock-based management is as much an exercise of institutional prerogative, in the last analysis, as it is a legislative or juridical mandate. Since 1998, the legal battle over Makah whaling—and, really, the bulk of what I have called the Makah whaling conflict, which NMFS has directly overseen—has been situated in the juridical space created by one key passage of the National Environmental Policy Act (NEPA), which states:

> [All] agencies of the federal government shall include in every recommendation or report on proposals for legislation and other major federal actions significantly affecting the quality of the human environment, a detailed statement on the environmental impact of the proposed action. (42 USC § 4332)

Nowhere is it expressly stated that regulators' concern for gray whales must be delimited to questions of populations and stocks. *Stock* is the way that NMFS has operationalized the category of "affected environment" in light of its legislative mandates and its own institutional history.

A full genealogy of these categories is beyond the scope of this book. What matters is that the moral classification of whales as killable—and the ontological classification of whales as irreducibly plural—is thus inherent in the way that the institution *sees* whales, to borrow James Scott's (1998) meta-

The Science Has Ruled / 143

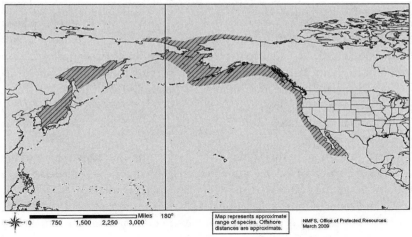

Figure 6.1 Gray whale stocks range, as estimated by NMFS Office of Protected Resources, 2009

phor for state measurement and control. Broadly speaking, NMFS evaluates proposals for indigenous whale hunts as it would any other "living marine resource," in terms of population estimates and theoretical net growth rates. The only significant difference is that for cetaceans, NMFS calculates a stock's potential biological removal (PBR) relative to its biological carrying capacity rather than to its *minimum* sustainable population. Still, if a proposed harvest is less than the PBR minus other expected human-induced mortalities (such as entanglement in fishing gear or ship strikes), then NMFS may approve the harvest at its discretion. The 1997 stock assessment report for the Eastern North Pacific gray whale, for instance, calculated the stock's PBR at 432, well above the five whales per year requested by the Makah Tribe even after subtracting other possible human-induced mortalities. On the basis of this single calculation, NMFS initially granted the Makah Tribe's request to hunt Eastern North Pacific gray whales.

The language and logics of the agency's management model borrow heavily from agronomic models, as does much of modern Euro-American wildlife management.[11] These models were originally based on scientific forestry methods developed in eighteenth-century Prussia and Saxony, and one can see in the map of gray whale stocks (figure 6.1) the ideal of neat and orderly representations of nature that reach their fullest expression in the remaking of forests into single-species row crops (like the red pine plantations in the United States and elsewhere).[12] The state cannot plant whales in rows—erstwhile boosters of commercial whaling long ago abandoned their dreams of whale farms built in Pacific atolls, which would have been the closest cetacean equivalent—so carving up ocean space using a two-dimensional cadastral-style map now represents the closest approximation of the forestry ideal. A fisheries biologist working for NMFS in one of its protected species offices once shared his frustration with me that fish are so unlike trees—spending most of their lives on the move, shielded from direct human observation, and prone to crossing the lines that the state has drawn around them.

One result of all of this is that there is very little conceptual space in the specialized dialect of federal cetacean management for the consideration of whales as anything other than numbers expressed in terms like *quota, annual yield,* and *optimum sustainable population*. As such, the framework precludes the very debate that activists want to have about whaling. It is difficult to even talk about individual whales in policy forums, let alone to suggest that killing whales is unethical. The word "kill" rarely appears in federal documents pertaining to whaling. One speaks instead of harvests, or of the number of "removals" a population will support in a given time period.

The fisheries management approach and its attendant register are, to borrow from Carol Cohn's analysis of technical registers in a different context (1987), transformative rather than additive, as together they actively exclude certain topics from consideration. Technically, NMFS does not prohibit anyone from arguing that whales should be treated as individuals or that whales should be given legal protections in excess of those of fish. Instead, it treats those objections as nonsensical and irrelevant, as outside of the bounds of relevant "conservation issues."[13] Margaret Owens explained:

> If you look at the comments submitted to NMFS [during the public comment period of the proposed rule changes], you can scroll through a great many comments saying you [shouldn't] kill beautiful, intelligent whales. NMFS never addresses these comments. They never say in a DEIS [draft environmental impact statement], 574 people commented that it is wrong to kill whales. . . . That's because the government doesn't ask, should whales be killed? They only ask, should *these* whales be killed. *How many* of these whales should be killed?

The distinction is an important one. Morally speaking, there is a world of difference between a blanket prohibition on killing whales and a harvest quota of zero. By the time I arrived at my field sites in 2010, Margaret and other activists had taken it for granted that they needed to argue exclusively for the latter.[14]

Stumbling Blocks

The transformation of the activists' tactics did not happen all at once. As the limits of roadside protest and direct intervention became apparent, Chuck and Margaret found themselves focusing more and more on generating what they referred to as "stumbling blocks," or strategic points of contestation that, if resolved, would do nothing to alter the group's fundamental opposition to whaling. When they were not holding roadside protests, PCPW was busy developing an encyclopedic knowledge of federal regulations to strategically use in their favor. During our conversations, Margaret would frequently disappear into the back room and bring back stacks of binders filled with newspaper clippings, articles from scientific journals, and government documents obtained through the Freedom of Information Act. She kept detailed journals of her actions and conversations.

"You don't even know half of the different issues that have come into this thing," Chuck said. "When there's something to go on, on our issue, we go on it. *We work it.* As best we can. And it's surprising what comes out of it. Anything connected to whaling, we went after it. It's just one thing after another."

The group argued, for instance, that the risk of stray bullets from the whalers' .50 caliber rifle presented an unacceptable hazard to other boats in the area and to people on the nearby shore. After learning about the weapon's range from a ballistics expert, they notified the local newspapers and radio stations and delivered information packets to county commissioners. In another case, the group learned that a Makah man who worked at a nearby state correctional facility had fashioned a metal whaling implement using equipment and materials at the facility. After state officials issued a statement declaring that they would no longer allow whaling equipment to be made from state property, PCPW "thanked" them with a demonstration outside the prison. The point, Chuck told me, was to "throw every hurdle" at the Makah whaling effort.

As the conflict continued, PCPW's stumbling blocks became more technical and bureaucratic. At first unaided by computers, Margaret compiled the necessary information through low-tech means, sending handwritten letters and using copy machines. A fellow activist drove from Ohio to Maryland copying "every document the National Marine Fisheries Service had on gray whales." When NMFS released a seven-hundred-page EIS on Makah whaling in 2008, Margaret drafted a line-by-line commentary and rebuttal. In her comments, she increasingly relied on the MMPA's ambiguous protection of US citizens' "aesthetic" right to the enjoyment of whales and dolphins.[15] Tami Drake, the paralegal who joined the activists on the water, carried an indexed compilation of state, federal, and Coast Guard regulations with her on the boat. It was her job to watch for any infraction by the Makah and report it immediately via her mobile phone.

"The thing we've learned," Chuck told me, "is you use the government's own words and science, you just use everything they put out, against them. Of all the things I've learned, I think that's the most important."

The immediate importance of the aforementioned study on whale population genetics (Lang et al. 2011) depends upon its role within a series of lawsuits filed by the opponents of Makah whaling in federal court beginning in 2000. It is useful to think of these lawsuits and the procedural objections on which they are based as "hurdles" as well. In the first lawsuit, *Metcalf v. Daley* (14 F.3d 1135 [9th Cir. 2000]), animal advocacy groups argued that NMFS had violated NEPA by officially approving a Makah hunt without first conducting a rigorous environmental assessment (a publicly recorded inquiry that determines whether a lengthier EIS is necessary) of the tribe's proposal, which is the minimum required step in the case of federal actions with possible environmental consequences. NMFS officials had, in fact, assembled an environmental assessment in 1997 after advocacy groups complained about the agency's failure to do so. The US Ninth Circuit Court ruled for the activists and expressly mandated that NMFS take a "hard look" at the environmental risks associated with the Makah whale hunt. In 2001, in compliance with the court's orders, NMFS completed a new environmental assessment that was mostly identical to the 1997 version. Activists responded with a second lawsuit, *Anderson v. Evans* (371 F.3d 475 [9th Cir. 2004]), this time calling for an EIS in light of concerns about the vulnerability of the resident/PCFG gray whale population. The plaintiffs (who included Chuck and Margaret) also argued that the Makah hunt should be subject to the Marine Mammal Protection Act. The court agreed on both counts, demanding a full EIS and ordering the tribe to seek a waiver of the MMPA before resuming the hunt.

The delays that followed from these lawsuits center around the resident/PCFG whales; and antiwhaling activists, especially Chuck and Margaret, were instrumental in bringing the subpopulation to the attention of federal regulators. Prior to 1997, in fact, the PCFG did not exist as such. People living on the Olympic Peninsula knew there were gray whales in Puget Sound during the summer, and as early as 1984, behavioral ecologist James Darling

Figure 6.2 PCPW antiwhaling advertisement, featuring "adopted" whales, that appeared in the *Peninsula Daily News*, May, 11, 2000
Courtesy of Margaret Owens

had used aerial surveys to estimate their population. But federal officials made no mention of the resident whales when they removed the Eastern North Pacific stock from the Endangered Species List in its entirety in 1994. Between 1997 and 2002, antiwhaling activists in the Makah whaling conflict brought attention to the resident whales with a series of protests and lobbying campaigns. Activists held up signs at roadside protests, promoted research they viewed as friendly to their interests, and circulated the names and images of gray whales they "adopted" through the Cascadia Research Collective. Chuck and Margaret placed a full-page advertisement in a local newspaper warning of the threat a Makah hunt would pose to the resident whales (see figure 6.2). Layering contemporary gray whale sighting data onto a nautical chart, Margaret created a map showing historical site fidelity by resident whales and sent it to NMFS. Eventually, NMFS was compelled to take notice. Mentions of the putative resident subpopulation in the agency's official stock assessments went from brief and dismissive to lengthy and substantive in the decade following *Anderson*, helped along by new advances in population genetics that showed subtle differences in mitochondrial DNA

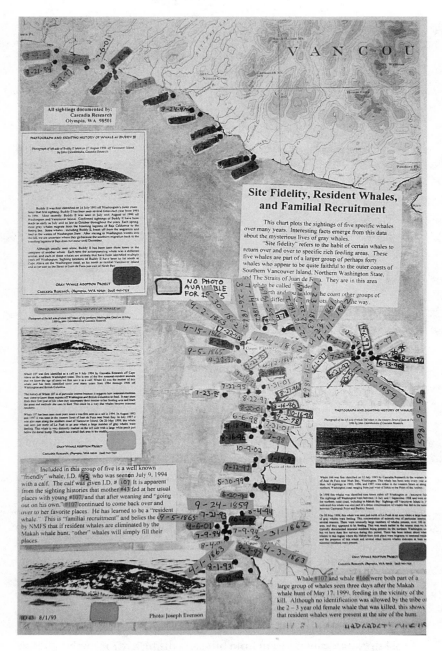

Figure 6.3 Homemade map by Margaret Owens showing gray whale site fidelity
Photo by the author

(mtDNA) haplotype frequency among the residents.[16] In 2008, two Canadian scientists announced they had found mtDNA differentiation among the PCFG (Frasier et al. 2011), findings the activists eagerly promoted in their comments on the draft EIS (DEIS) the same year. In his capacity as the biologist for the marine mammal division of the Makah Tribe's fisheries department, Jonathan Scordino continued to push back against these trends, as he encouraged "caution in the interpretation of the genetics" and noted that there were "caveats to interpretation such as high genetic diversity of the PCFG" (Scordino, Bickham, Brandon, and Akmajian 2014). In 2012, NMFS announced it would discard the previous DEIS in light of the findings in Lang et al. (2011), which replicated the earlier findings by the Canadian scientists. This was the "science" to which Margaret had excitedly referred me when we first met, and it is responsible for the current hiatus in Makah whaling at the time of this writing.[17]

In 2013, NMFS convened a task force on gray whale stock identification with the primary purpose of providing advice on whether the PCFG ought to be considered a separate stock (Weller et al. 2013). Activists had done nothing less than push NMFS to rethink how it defined the parameters of a cetacean management unit.[18]

You Can't Tell Them Apart

While these legal delays and regulatory second-guessing have been a boon for antiwhaling activists, it is slightly less clear how these new developments will support the activists' broader goal of speaking for *all* gray whales. Margaret's persistence on these lawsuits has even caused confusion among other activists, who have asked whether it means she and Chuck are okay with people killing gray whales as long as they aren't killing *resident* gray whales. Margaret's response is always "No, but we *really* don't want Makah hunters killing resident whales."

Margaret does her best to distinguish PCPW's motives from its tactics. "All the binders [full of data], all the avenues of pursuit that we've put our tentacles into," she told me insistently, "none of that had anything to do with our decision to get involved. None of that had to do with just the gut knowing/feeling [that] you don't kill whales. These are just the things we had to learn, we had to track down to argue. Because you can't write a comment to the government saying, 'You just don't kill whales, it's just not right.'"

Despite her efforts, over the last ten years the line between some of the group's tactics and motives has begun to blur. When it comes to "the science," Margaret and her group seem to have abandoned that distinction entirely. There is no question that the findings in the Lang study ushered in a major tactical victory for antiwhaling activists, as it delayed the completion of the court-mandated EIS and placed serious obstacles in the path of the Makah management plan. The strategy relies on the inherent uncertainties of population modeling in conservation biology, the ultimate irreducibility of which offers activists the seductive promise of perpetually renewable delays.[19] In the meantime, the quota for gray whale "removals" in the area around the Makah usual and accustomed fishing grounds is effectively zero. What is striking, however, is how Margaret and other activists seem to present the scientific findings as if they were the final word on the legitimacy of Makah whaling. In fact, the Lang study derives its significance from the conservation-oriented, bureaucratic system of marine mammal management implemented by the National Marine Fisheries Service. One way to interpret Margaret's declaration that "the science has ruled" would be to assume that the findings had somehow affirmed the activists' ethical position or at least vindicated the assertion that killing Eastern North Pacific gray whales is categorically imprudent. None of this is the case.

"The crucial question," the court wrote in *Anderson*, "is whether the hunting, striking, and taking of whales from this small group could significantly affect the environment in the local area." Even here, in the context of this major legal victory for activists, whales exist only as component elements of the natural environment or objects of aesthetic appreciation.

Antiwhaling activists' efforts on the issue of Makah whaling are now almost entirely concerned with the NMFS EIS and MMPA waiver process. Activists write comments on new drafts of the EIS. They greet new scientific findings on stocks and genetics and populations with eagerness. They are, so to speak, all in on "the science." So far, this strategy has brought them success. The tactics that PCPW refers to as "stumbling blocks" have helped slow the administrative reapproval process of whaling to a virtual halt. Having accepted the form, the activists have won on content, able to eke out temporary victories against an organization (the Makah Tribe) with a larger budget and much greater familiarity with the fisheries management system. By adjusting their tactics and objectives to fit within the federal management paradigm, however, they have also allowed the discussion to move away from the ethics of whaling, toward a discursive space where the killing of whales is tacitly acceptable, if momentarily imprudent.[20]

Sandra Abels, an activist who opposes the killing of individual whales on moral grounds, begins her public comment on the 2012 DEIS with the following:

> I strongly encourage NOAA/NMFS to not allow the Makah a waiver to hunt whales at all. There is NO way to determine which gray whales are from the main Eastern Pacific population, the residents or the highly endangered Western Pacific population. Killing a whale from the resident whale or Western Pacific population could prove catastrophic to those populations. They don't wear name tags. There is no way to tell them apart!!

One could add to this the following implication: if Makah hunters *could* reliably tell the two stocks apart, it would be acceptable to hunt them. This is not what Abels intends to say, but such is the subtle power of bureaucratic authority. Effective bureaucracies compel potential dissidents to repeat their own implicit moral frameworks back to them.

Given that the Makah Tribe has been disallowed from whaling for over fifteen years as a direct result of activist efforts, it may seem counterintuitive to suggest that the federal management system is configured against the activists. And indeed, activist victories may continue indefinitely. If Makah hunters find a way to reliably differentiate migratory whales from resident whales—and several possibilities have been discussed, including an exhaustive identification catalog based on individual gray whale skin markings—activists may simply come up with another hurdle to delay the reapproval process. The point, however, is that activist victories can only ever be temporary. The activist lawsuits that have stalled Makah whaling could very easily end with a renewed quota and a green light for Makah whaling. But they could never end with a permanent ban on hunting gray whales. The default outcome, given the current structure of US law, will always be federal approval for a tribal whale hunt. In the meantime, the moral economy of NMFS requires that activists affirm what is arguably the inverse of their moral principles in order to have a figurative seat at the table.[21] The recent history of activism in the Makah whaling conflict shows how difficult it is for activists to sustain such tactics without tacitly accepting the moral assumptions of the very systems they are attempting to change.

SEVEN

The Whale Approaches

For all the salience of whales and whaling in the Makah community, for all the pervasiveness of whale-as-symbol, and for all the years in which the figure of "the whale" was at the ostensible center of debates about Makah treaty rights and tribal sovereignty, it is striking how infrequently actual, living gray whales become visible in daily life for the residents of Neah Bay. I did not see a single whale during my first eight months on the Makah Reservation, even though I was there to study whaling, and even though I spent an inordinate amount of time hanging around fishermen and former whalers. To "see" gray whales from shore means spotting tall spouts of water vapor on the distant horizon; sharp eyes or powerful lenses might catch glimpses of dark little slivers just above the waterline as the whales skim the surface before diving again to feed. Makah commercial fishermen get closer looks, but fishermen more often encounter sperm whales and other fish-eating cetaceans.[1] And when gray whales show up in the speech of Makah whalers, it is usually not as an acting subject but rather in the form of the idiom "a whale on the beach," uttered wistfully by whalers in dreaming of a successful hunt. After just a few months in Neah Bay during which no whale hunts took place, I noticed that I, too, was beginning to see the prototypical form of gray whales as "on the beach."

Even if one is fortunate enough to see gray whales up close, they may do little to acquit themselves as magnificent creatures. A federal marine biologist once called them "swimming rocks." Admirers try to find nice things to say about their acrid breath, which some say smells like a mix of sulfur dioxide and rotten mud.[2] Some activists may have developed a naturalist's appreciation for gray whales' behavioral repertoire of spy-hops, lobtails, and slapping, but rarely do gray whales do anything so spectacular as breach, as we are accustomed to seeing from orcas and humpback whales. Gray whales are not even first on the list of whales that Makahs would like to eat. That honor goes to the humpback whale, which is featured in the Makah Whaling Commission's insignia and which, as indicated by archaeological records, was consumed with slightly greater frequency than gray whales prior to contact with Europeans.[3]

But there is one thing gray whales do consistently that resonates in human communities: they approach. In the gray whale birthing lagoons of Baja California, gray whale mothers approach boats filled with first-world tourists and lift their newborn calves above the water to be touched and petted; the tourists, in turn, are made delirious with excitement (Siebert 2009). Transformative experiences at the lagoons helped inspire resistance to the construction of a major salt works there in the 1970s and set more than a few activists on their way to Neah Bay in the 1990s after the Makah Tribe announced its intent to hunt gray whales.[4] The first Makah hunt was not successful until a gray whale approached the canoe.

This chapter examines how different parties to the Makah whaling conflict make sense of this most interpreted of gray whale behaviors: the whale approaches.

"It Chose Us"

It was what Wayne Johnson would later describe as a beautiful day in early September when he, Theron Parker, and three other Makah men, armed with harpoons and high-powered rifles, boarded two powerboats, cruised several miles east of the Makah Reservation along the Strait of Juan de Fuca, and attempted to kill a gray whale without written sanction from the tribe or approval from the US federal government.

Eight years had passed since the historic whale hunt of 1999. The family hunts of 2000 had been unsuccessful, leaving less of a mark in the community's collective memory, and a federal court had ordered a halt to Makah whaling soon after. In the early part of the decade, newly elected tribal council members had questioned the emphasis the previous council had placed on whaling, citing other pressing needs in the community. The point, they said, had simply been to protect the treaty right, and that had been accomplished. Keith Johnson, still in his capacity as president of the Makah Whaling Commission, told the *Seattle Times* in 2002 that whaling was a "dead horse" (Mapes 2002). By the middle of the decade, however, a new tribal council came to power that was more sympathetic to whaling. Tribal leaders began secretly discussing the possibility of "whaling on [their] treaty rights." They felt an unauthorized hunt might force the courts to rule once and for all whether the Marine Mammal Protection Act abrogated the 1855 Treaty of Neah Bay. They found a sympathetic ear in a few key members from the 1999 crew, especially those with checkered pasts who had seen their stars rise in the buildup to the first hunt.

Wayne had worked for a while as a truck driver after whaling was suspended, battling alcoholism and struggling to maintain consistent employment. "There were a lot of council meetings [and] whaling commission meetings," Wayne recalls of the time leading up to the rogue hunt, "and three and four [tribal] council members kept saying [that] we should just go get [a whale]. And that's like a hint, you hear that over and over again, you start thinking they're saying, 'Let's go get one.'"

Theron had achieved something like celebrity status as the harpooner in the 1999 hunt. "My kids may never be able to meet Michael Jordan," one Makah man beamed in a television interview after the 1999 hunt, "but they can *touch* Theron Parker."

Theron remembers a similar but more direct message from the tribal council in the months leading up to the rogue hunt:

> We wanted to go get another one. [A strong supporter of whaling] was in [the tribal] council, and he was part of the whaling commission, too. And being, wearing both hats at the same time, or maybe not thinking—I don't know what the whole deal is . . . he said, "Well then, *go get the goddamn thing*." I told him, "Well, I can't right now 'cause I'm busy working on a boat, but September would be ideal for us, we could just go right out front here [in the Straits], 'cause they'll be all around. . . . We can just get it, stick it, bring it to the beach, and we're fed." So, that's when he said, "Well, just go get the goddamn thing."

Wayne and Theron offered me accounts of the hunt that differed sharply in tone and emphasis, befitting the two somewhat different approaches to whaling within the community that were reflected in the 1999 crew.

Wayne's narrative begins with a focus on the more practical aspects of preparing for the hunt. "We [got] a crew ready," he began, "laid out the guns, some paper to wrap the meat up in. We had a thirteen-foot boat and we had one bigger boat. . . . So we went out, and because we were in a little boat we decided—alright, we'll get one in the Straits."

Wayne would later tell a local reporter that the whale approached the boat before it was harpooned, but curiously, he leaves that out of his account now:

> We went around the corner—went to look at Sail Rock. Sure enough, there were two there in the kelp. So we got a harpoon in it, Theron got the first

one, and it went down, tried to dive down. But [the water] wasn't too deep so it couldn't go too far down. . . . We wanted seven harpoons in it before we even tried to shoot it. So, it towed us for a while—it was a big one, forty-five feet. We had the .577 [whaling rifle] on the bow of the boat. [Another crew member] went and picked it up, but it went off in his hand, shot a hole in the boat above the waterline. [The gun] came flying back, thirty-pound gun hit me right in the face.

The rifle flipped over the edge of the boat and sank to the bottom of the strait. One of the men on the other boat had a smaller rifle ("the .466"). But "he kept shooting it in the nose," Wayne explained. "He didn't know the biology of it, didn't know where to shoot it." By the time the Coast Guard arrived, the whale had been "knocked unconscious" by the shots and was tied up to Wayne's boat.

"It was breathing out blood, five or six feet high out of its blowhole," Wayne said, drawing in and exhaling a deep, slow breath to imitate the sound of the injured whale. One of the hunters, he said, was pleading with the Coast Guard: "Can we dispatch this thing?! Will you let us dispatch this thing?!"

With his cell phone, Wayne called a friend onshore and asked him to bring out another man who had been trained to use the gun. But the Coast Guard intercepted the shooter and took all five of the hunters into custody. In jail hours later, Wayne and the others were told that the whale was still alive. It later died, nearly ten hours after it had first been harpooned and shot, and it sank to the bottom of the sea.

Theron's account of the hunt begins, in contrast to Wayne's, with an emphasis on the spiritual aspect of whaling and the importance of proper preparation for the hunt:

> September rolled around, here come [another crew member] over to my canoe carvin' shed, and he says, "Let's go get a whale." I'm like, "Oh. *Oh.* September, yeah. Um. Let me go get ready." So I got my equipment out and did my thing, ya know, doing my bathing or however to get myself spiritually in *tune*. And I said, "Give me a few days." So, a few days came up and I said, "I'm ready."

Theron speaks quickly, and his syntax is coarse, with a resonant voice that sounds deeper than it is. Robert Sullivan (2000) described Theron as "menacing" and admitted that Theron frightened him (192). I found him somewhat threatening as well, but I think that was because rumors had been circulating about his violent temper. In our interactions he was always humble and courteous, and he is calm, almost soothing, when he describes the rogue hunt:

> So, we did everything like we were supposed to, we went and checked 'em all out, [and] I said, "Let's get a big one this time. Let's get a real big one." So, we went and checked 'em all out and—this—one—kinda—presented itself, I guess you'd say. . . . It come up and it was lookin' at me, I'm lookin' at him, I said, "I want that one right there." So I told the guys on the other boat, "Stop your motor . . . he'll come here." So it came right to us, and that's when it started. We—we got him.

Most striking, perhaps, is that Theron's account seems to lack any sense of violence or struggle.

> And, y'know, [we] had . . . everything all under control, and we had [the whale] coming like—it was comin' right into the bay, before the Coast Guard came. I got songs that come to me, you know, from, through doin' all my prayer, and stuff, and so I just kinda sang it, just like, like a horse on a bridle I just pull, a little bit on one line and it just turned, and it just, everything was just nice 'n smooth, it's goin' with the tide, just flick its tail and just keep—we had it. We had it, it knew it was beat, we had it. Brought it—we were bringin' it—to the beach, then the Coast Guard came racin' out there and got everything all excited and got it all twisted and . . . [the] thing, it went crazy again.

The version of the story that Theron shared with me ended there, before the Coast Guard apprehended the men and took them into custody.

The difference between Theron's and Wayne's accounts mirrors the men's position as rivals within the community and in media accounts of the 1999 hunt. Wayne's account focuses on the technical details of the hunt, and he does not spare some of the harder facts of the botched kill. In this aspect, his narrative fits with the views of the group that some in the community call the "bad boys," a mix of fishermen and administrators closely associated with Wayne who distanced themselves from talk of spirituality, professed to think of whales as no different from giant salmon, and were more likely to pose with pictures of guns than harpoons or ancient whaling gear. Theron represents, on the other hand, a suite of views and practices more common among his extended family, the descendants of Wilson Parker, who take a more outwardly spiritual view of the hunt.

Janine Bowechop, as the director of the Makah Cultural and Research Center, supported Theron over Wayne in the buildup to and narratives of the hunt. She was disappointed when Wayne was given the starring role in Robert Sullivan's account of the hunt and was embarrassed by Wayne's many public verbal miscues, as well as his lapses into violence and drug use.

Wayne's supporters are aware of his struggles as a public figure, but they choose to focus on his hereditary rights. "Wayne has been down and out, but no one can dispute that he came from whalers," Keith Johnson told me when I asked about Wayne's involvement in the hunt. "No one can question that he came from a whaling family."

Theron himself was conciliatory toward Wayne when he spoke to me about him in 2012. "There was a little bit of controversy between me and Wayne," he said. "The funny thing is, without [Wayne], for some reason, it just didn't work. You gotta have this mesh of people."

There are many viewpoints and versions of traditional spirituality in the Makah community that lie around or between Theron's immersive spiritualism and Wayne's stated lack of interest in "the spiritual stuff," but in the aftermath of the rogue hunt Wayne chose to remove this distance. When asked about the hunt by a reporter in 2007, he said a curious thing of the whale: "It chose us."

Why would Wayne say this? I was never bold enough to ask him about it directly. It was the first and only time I heard him refer to the idea that gray whales offer themselves to hunters. This is the man who took every opportunity to tell me that he "didn't get into the spiritual stuff." Nothing in his daily conduct or the way he went about preparing for a whale hunt suggested otherwise.

Perhaps Wayne really was attempting to make sense of the whale's behavior. Regardless, he did not have to reach far for a readily available discourse that packed a potent political punch. In its furthest reaches, the idea of gray whales offering themselves to hunters has the potential to chill speech by refiguring all contrary claims as assaults on religious exercise. But more immediately, in a debate that for most Makahs is as much about tribal sovereignty as anything else, it stakes a claim over the very meaning of *the whale*.

Two Senses of Sacrifice

Amid the talk of whales offering themselves to hunters, two distinct senses of the expression "the whale offered itself" seem to emerge. The first sense is an interpretation of concrete whale behavior: the whale in the 1999 hunt turning to "receive" the harpoon, Theron watching the whale "come right at" the boat during the rogue hunt, or Wayne saying the whale "chose" them. In each of these cases, it is possible to imagine the whales behaving differently. They might have kept their distance or swum in the opposite direction or lashed out violently at the boat. But they did none of these things. They approached, and in approaching they were perceived to have signaled their offering.

The second sense of gray whale self-sacrifice is less attached to observable whale behavior. Nuu-chah-nulth scholar Charlotte Coté opens her book *Spirits of Our Whaling Ancestors* with the following passage:

> It was May 17, 1999, the day my sister Charlene called . . . to share the news that members of the Makah Nation had been successful in their whale hunt. . . . A thirty-foot *maa'ak* (gray whale) gave its life to feed the Makah people, an act that elicited in me a sense of excited disbelief. (2010, 3)

Coté was in California at the time, so it is not likely that she knew the details of the whale's behavior. Here, "the whale gave its life" is just another way of saying, "the hunt was successful."

Coté infers that the whale offered itself from the fact that the whale died at the hand of the hunter. The whale's death is, by itself, a necessary and sufficient indication of its sacrifice. This is why some Makah whalers see no difference between a whale that is killed in a hunt and a so-called drift whale that washes up on the beach with its blubber still able to be harvested. Historically, Makah and Nuu-chah-nulth hunters viewed both kinds of events as an answer to prayer. When Theron Parker told me about the whale he and his crew butchered after it became beached in the Yellow Banks area of the Olympic National Park in 2001, he referred to it simply as "the second whale," pointedly making no distinction between it and the whale he harpooned in the 1999 hunt.

Greig Arnold invoked a similar epistemology when I asked him if he thought the whale in the rogue hunt offered itself to the whalers: "Did it come home?" he replied, asking rhetorically. "No? It didn't give its life. It fought death to the end."

I asked Greig about the old days and the stories of whales towing whalers out to sea for several days. "Someone could look at that," I said, "and say the whale is trying to get away, it isn't giving itself. Can it *look like* the whale is running away, and still you would say the whale is giving itself?"

"No," he answered. "[If] it's running away, you ain't getting it. You ever try to approach a whale on a boat? It's not easy. It's really not easy. I don't know what kind of intelligence you think they have or don't have, but it isn't about what they *don't* know. They know they're being hunted. They don't stick around. You just have to find one that is like, OK, this is the moment." For Greig, this second sense of gray whale self-sacrifice is based on an inference: the whale is too swift, too intelligent; therefore it cannot be captured unless it willingly allows itself to be.

Theron's account of the rogue hunt clarifies the relationship between these two senses of cetacean self-sacrifice. For Theron and others like him, there is, ultimately, no distinction between the two; the first sense (the whale physically approaching) is a direct expression of the second. Consider the emphasis that Theron places on preparation for the hunt, which I first described at length in chapter 3. In the two years leading up to the 1999 hunt, much was made in the community of whether the whalers were preparing themselves properly. Were they purifying themselves, abstaining from drugs, alcohol, and sex? Were they talking to the elders about how to bathe in icy creeks and sing sacred songs? Almost everyone on the crew engaged in some kind of spiritual preparation, from bathing in creeks to beating themselves with stinging nettles to spending time in traditional sweat lodges. Many of the whalers with whom I spoke, however, viewed spiritual preparation as a practical precaution, a rational measure unconnected to any idea of gray whale self-sacrifice. For them, it made sense to prepare oneself physically and mentally to face the inherent danger of hunting a huge and powerful animal from a small and relatively fragile canoe. But they were reticent to say that they believed animals had spirits and offered themselves to hunters. For Theron, however, the end result of the whale offering itself is not easily divorced from other aspects of spiritual preparation. To prepare oneself for the hunt is an index of one's worthiness, and the whale's self-sacrifice is thus the product of adequate preparation.

Recall Theron's description of the somatic phenomenology of his spiritual preparation: the quickened pulse, the sharp sensory acuity, the feeling like you have "eight Red Bull [energy drinks] inside of you." All of this reaches a new peak when the hunt begins and a whale is finally sighted. Suppose now that a whale approaches the boat in that moment. It is here that the two senses of gray whale self-sacrifice merge. The whale's approach cannot be an expression of just any cetacean desire. It is expected as a result of prayer and preparation and proper conduct. It is an indication of the whale's sacrifice. The ultimate proof, however, as Greig explained, is not whether the whale physically approaches but whether the whale comes home to the beach at the end of the hunt. The whale's visible behavior is not necessary.

It is just one of many aspects of the hunt as a totality and can only be understood as part of this total social fact.

A Gift Not Freely Given

What would it mean for the whale to act differently, to choose not to come home? When Theron described the whale's behavior in the rogue hunt as "like a horse on a bridle," he also said a curious thing: "We had it, *it knew it was beat.*" Theron appears to reference a struggle that, according to his own narrative, never took place. In what sense was the whale, who approached the boat and whose behavior Theron described in such passive terms, defeated?

Despite the voluntarism that Makahs impute to whales who choose to offer themselves to human hunters, the cultural logics of Makah spirituality are actually rather deterministic in the metaphysical sense. One Makah elder told me this in nearly so many words, noting that Makah whalers have "always had a victory in the spirit realm before they went in[to] the physical." Theron's reference to the whale being "beat" implies something similar. If the hunt is conducted properly, the whale *must* acquiesce and "come home." Scholars have recognized a similar tension between talk of reciprocity and talk of domination in the attitudes of other North American hunting peoples (Brightman 1993; Tanner 1979). Although hunters frequently speak of animals "offering themselves," at times they also suggest an adversarial relationship between hunter and hunted. Nadasdy (2007) attempts to resolve this tension by framing the animal's offering as a gift, with all the associated obligations and expectations that come with it in what anthropologists have termed "gift economies." Not all gifts, Nadasdy reminds us, are freely given (28). Under this view, by conducting the proper rituals and promising to "take care" of the whale's spirit, Makah hunters oblige the whale's spirit to offer itself to them. This is the sense in which the whale is defeated. In essence, the concept of gray whale self-sacrifice ascribes such capabilities to the whale as to simultaneously grant and obliterate its capacity to choose.

A Protest Hunt

Theron's account of the rogue hunt makes it clear that despite its rushed appearance, the hunt was deeply meaningful to him. One of his spiritual mentors told him that it would take him less and less time for him to prepare for a whale hunt as he became more experienced, that it would be easier for him to reach what he called "that spiritual space" where one has the strength to do something as difficult as hunting a whale. "And it's true," he said. "I was ready within a few days."

The fact that the rogue hunt was believed in the community to have been conducted without proper spiritual preparation perhaps only shows the variability and individualism inherent in Makah spiritual practices pertaining to whaling. For Theron, it was perfectly regular. In the federal court proceedings after the rogue hunt, Theron pleaded down his charge to a misdemeanor, but Wayne Johnson and Andy Noel decided to stand trial. Their attorneys intended to use religious freedom as a defense, suggesting that Wayne and Andy were exercising their religious liberty in the act of hunting the whale. The judge barred the defense in a pretrial ruling, however, stating that it was not an issue of religious freedom because this had been a "protest hunt" (Coté 2010, 187–88).

The rogue hunt was widely considered a failure in the community not because it was a protest but because it failed *as protest*, setting back efforts to regain federal regulatory approval by several years or more. At the time, Makah officials were rumored to have been closing in on a deal for a congressional bill that would have overridden the judiciary delay. The rogue hunt scrambled that opportunity. The same officials who were pursuing the legislative fix might have been holding their breath to see if the rogue hunt worked, and according to Theron, some of them had even given their approval with a wink and a nod (so to speak). If it had been successful, this gambit would have proven the strength of the treaty under US law and set the tone for future negotiation.

One wonders what might have happened if the rogue hunt had not been botched, if the primary gun had not been lost and the shooter had known the whale's anatomy better. Would the whale have been confiscated even if the whalers had made it back to the beach before the Coast Guard responded? What would the authorities have done if the whale had been surrounded by a throng of excited onlookers in Neah Bay who were not keen to allow it to be seized? Wayne Johnson was always fascinated by the possibility of killing a whale during the annual Makah Days festival, when Natives from around the Northwest Coast gathered on the Makah Reservation for a weekend of festivities. "Let's see them take away the whale," he mused, "when it's on the beach surrounded by fifteen hundred Natives."

This was not on Theron's mind when he was preparing for and executing the hunt, however. "It wasn't a game to me," he said in discussing his feelings after tribal leaders disavowed the hunt and "left [him] out there to fry" with the rest of the crew. The sense of betrayal he felt over the tribe's handling of the rogue hunt seemed to me as much about the politicization of the hunt as the actual betrayal. "I want it out there," he said. "They told me to go get [a whale], I went and did what they asked, and then they just said they don't condone what I did. What am I supposed to do? How am I supposed to feel? You know? . . . I was angry at first but now it's just like I'm really hurt about the whole thing. It's just like, it was a big game. To me [whaling] wasn't a game."

"These Are Not the Same Whales"

Gray whales' tendency to approach humans is as salient for the antiwhaling activists and opponents of Makah whaling as it for Makah whalers. It perhaps goes without saying that the former interpret the behavior very differently.

The remote gray whale birthing lagoons of Laguna San Ignacio and Guerrero Negro in Baja California are premier destinations for whale enthusiasts in large part because of the tendency of gray whale mothers in the lagoons to approach boats and "present" their calves to awed tourists. In an article that appeared in the *New York Times Magazine* in 2009, Charles Siebert describes his experience with a friendly gray whale:

> Within moments, the mother was surfacing again off to our stern and doubling back in our direction, but this time with her newborn male in tow. . . . The baby gray glided up to the boat's edge, and then the whole of his long, hornbill-shaped head was rising up out of the water directly beside me, a huge, ovoid eye slowly opening to take me in. I'd never felt so beheld in my life. . . . I read before my journey to Baja of what happens to people when they come in contact with a whale, how they tend to go, literally and figuratively, a bit overboard: nearly tipping over boats for a passing touch; spontaneously breaking into song; crying out in ecstasy; or just flat-out crying. [Toni] Frohoff [a marine mammal behaviorist along for the ride] warned me as we were first boarding [the boat] that morning that she was given to doffing her scientist hat in the presence of a whale, and sure enough, there was . . . Frohoff dangling far out over the boat's prow, arms outstretched, cooing and trilling at the approaching mother and calf. Another watcher in our boat began singing Broadway show tunes. I joined in.

In the late 1990s the lagoons developed a kind of gravitational pull among the community of activists opposing the Makah hunt, a place where "hardened businessmen were routinely reduced to childish glee, and serious sci-

Figure 7.1 Makah elder Alberta Thompson touching a gray whale calf in Baja California
Photo by Jared Blumenfeld

entists dissolved into wonderment and uncontrolled laughter," as Joshua Horwitz describes the encounters.[5] Alberta Thompson claimed that her experience in Baja reshaped her understanding of whales and altered her motivation for opposing the hunt. In an earlier interview with a reporter in 1998, Thompson repeated a local saying about the relative gustatory quality of gray whale meat, noting that "gray whale is hamburger, and humpie is steak" (Andersen 1998). All of this changed when Thompson was approached by a mother gray whale and calf in the lagoons. After the encounter, Thompson said she felt she had developed a special connection with the whales, and that connection animated her opposition to whaling. "That's when she really fell in love with them," Thompson's daughter later told me. "That's when she knew she had to fight for them." When Makah hunters eventually killed a gray whale in 1999, news cameras captured Thompson as she wept and told reporters, "They killed one of my friends."

Antiwhaling activists reject the possibility that gray whales might offer themselves to human hunters. They generally look upon such claims with cynicism. "That's like something that's said to shut white people up, sorry," Margaret Owens once told me, "because how do you argue with . . . deep-seated religious beliefs"—she paused, choosing her words carefully—"that animals are *giving* themselves to torturous death by harpoon, gun, drowning? That's something that's said that you're not supposed to come back at. That's supposed to shut you up." Jonathan Paul was less diplomatic: "The whale's not gonna give itself to a bunch of humans to die for them," he said to me in an interview, arguing that it runs against everything that is scientifically known about animals. "That's a bunch of bullshit. That's a bunch of religious, anthropocentric, arrogant bullshit. Animals are not gonna come up and wanna give themselves to humans."

For activists, there was a straightforward explanation for the whale's decision to approach the canoe during the 1999 Makah hunt. "It spy-hopped them," Jonathan said plainly. "It just wanted to be petted," said another activist.

Some antiwhaling activists feared that the 1999 whale was a particular calf that they had come to know during a visit to Baja two years earlier. That turned out not to be the case, but many activists were openly worried that their encounters in Baja would make the whales easier targets for Makah whalers. Sandra, one of the leaders of an organization that promotes shore-based whale watching in the area, felt those changes had already taken place. She visited the lagoons in 1998 and was able to "look into the eye" of a gray whale. "Whales change," she told me earnestly. "These are not the same whales that [the Makah whalers'] grandfathers were hunting."

These are not the same whales. The statement is notable for its clear implication that human agency (tourism in Baja) might be reshaping the nature of nonhuman animals (the "friendly" gray whales) who then approach other humans at a different time and place and become subject to competing interpretations all over again. These are not the same whales, yet the activists' greatest fear comes from their commonsense intuition that whales remain the same as they cross ocean space and pass through different zones of human influence. The most tragic possibility of all, for activists, is that these *are* the same whales and that the gray whales' "friendliness" is not a new phenomenon. Makah elders tell stories of whales offering themselves to whalers many generations ago. Makah fishermen tell more recent stories of whales approaching fishing boats night after night, each night drawing closer until finally they peer over the rail and fix a giant ovoid eye upon the boat's occupants. Could these be the same behaviors? What if whales have been trying to "make contact" for hundreds of years, and humans everywhere have responded by killing them? This, above all else, was what the activists feared.

"Gray Whales Are Large, Wild Animals"

The official records of the National Marine Fisheries Service, including the most recent draft EIS, serve as a record of institutional memory, and nowhere in the one-thousand-plus pages of the statement does it refer to the whale approaching the boat in 1999 or 2007. There *are* numerous references throughout the document to "approaches" as a countable noun, but always as an action performed by the whalers in relation to an implicitly passive quarry.[6] The reports seem to go to great lengths to *avoid* describing the whale's behavior in the 1999 hunt in so many words, stating that the whale "surfaced on the right side of the canoe, and the crew harpooned it as it moved across the bow . . ." (1-37–38; quoting Gosho 1999). The agency's account of the rogue hunt similarly states: "Sometime on the morning of September 8, the hunters approached a gray whale approximately 40 feet (12.2 meters) long near Seal Rock and harpooned it" (1-40).

Regarding the more general propensity for gray whales to approach boats, NMFS has officially commented on this issue only once in the context of the Makah whaling conflict. Summarizing available research on the behavioral effects of whale watching, the EIS acknowledges that "some observers noted that some gray whales were attracted . . . to quiet, idling, anchored, or slow-moving vessels" (NMFS 2015a, 3-189). Parsimony may be the goal here, but note that the word *attract*, with its metaphoric associations to the behavior of charged particles in magnetic fields, is perhaps the most tested way of substituting passivity for agency in describing moments of encounter. Responding to the concerns of Puget Sound whale-watching guides that a Makah hunt might cause whales to be warier of boats and, thus, "less approachable" (NMFS 1997, 39), the agency's 1997 environmental assessment states:

> While the behavior of individual whales near boats might well be affected if they are wounded but not killed by Makah hunting, it seems unlikely that this would change the behavior of other gray whales. . . . NMFS acknowledges the concern that wounded whales could be dangerous. This is true, as it is for any large animal in pain. However, there is no obvious reason why a wounded animal would seek out innocent vessels upon which to wreak destruction. It seems much more likely that a whale once wounded by a boat

would avoid other vessels. In any case, gray whales are large, wild animals, and persons should exercise caution in approaching any whale, regardless of its condition. (39)

This comment, already notable for being one of the few instances where the agency mentions whales as individual living beings rather than as constituent elements of the "affected environment," does more than simply chastise the whale-watching industry for failing to show sufficient scientific detachment, although it certainly does that in its final, exasperated note of caution.[7] This is, after all, an industry that is already suspect from the point of view of conservation management for having based its entire business model on a behavior so disruptive of the boundary between *human* and *wild*. The comment takes aim at the very idea of amiable human-whale interaction, reminding would-be whale watchers that these kinds of encounters are not only illicit but also dangerous. By officially cautioning humans against approaching whales because they are large, dangerous, and unpredictable creatures, NMFS imprisons whales in their wildness, refusing them any ground to disrupt the species boundary by initiating contact with humans.

What about the Whales?

I have spent most of this book exploring the different enactments of gray whales in the Makah whaling conflict—as food, as spiritual beings, as majestic others, as elements of population models. We have examined the political stakes of the conflict, and we have seen how the consonance and dissonance of the moral economies of different parties can span ontological divides, helping to give the Makah whaling conflict its shape and going a long way toward determining its future. What about the whales themselves? Are they actors in this story? Can the whale be more than the inscrutable surface upon which human interests are inscribed?

Anthropologists start out at a disadvantage in this regard. By attending first to what our human informants say and do within the context of their relationships with animals, we are always already a step removed from the animal subject (see Marvin 2010). We must, in other words, deal with multiple layers of erasure by interpretation. As this chapter has shown, even the whales' approach is erased to some degree, used to bolster existing cultural narratives. Makah whalers use the behavior to support their belief that whales offer themselves to hunters. Activists use the behavior to support their portrayal of whales as intelligent, friendly, and curious others.

Scientists still know very little about the structure and functioning of gray whale brains, and even less about gray whale minds.[8] They do not know if the act of approaching humans is even among a gray whale's most exciting or salient experiences, assuming that excitement or salience are categories that apply to gray whales' mental life. Gray whales' suspenseful annual sprint past the swarms of killer whales that hide in the deep canyons of Monterey Bay[9] or any number of events humans neither witness nor fully understand could be far more important to the whales.

For their part, the gray whales that were in the area during the 1999 Makah whale hunt did not appear to be terribly interested or involved in the human drama that was playing out around them, from the news helicopter circling overhead to the canoe full of whalers that paddled toward the whales every time they surfaced. If the whales knew they were being hunted, as Greig and Theron both argued, they did not appear to consider the hunters much of a threat. The many hours of aerial news footage of the hunt provide a repository of whale-as-prey behavior, and at no point did the whales appear to be actively avoiding the hunters. From an outside (and overhead) perspective, the whales seemed difficult to kill because they were constantly moving as they foraged for food. In fact, a parsimonious description of the footage might go something like this:

> Gray whale surfaces, mud streaming from its mouth.
> Whalers paddle furiously toward the whale.
> While canoe is en route, whale dives.
> Repeat.

It wasn't until a juvenile female approached and lingered near the canoe for several minutes that the hunt was successful.

Gray Whale Agency

Influential theories of nonhuman agency in anthropology have tended to flatten the distinction between the agency of animate and of inanimate things, ignoring intentionality and treating the different "actants" similarly (e.g., Callon 1986; Latour 2014).[10] Jane Bennett's (2009) notion of "vibrant matter," for instance, makes no clear distinction between the agency of gunpowder residue and that of a stick of wood, just as Latour makes no distinction between the agency of farmers and that of microbes (1983), of armies, and of riverbanks (2014). By these definitions, gray whales are clearly important agents, or "actants," in this story. The oil content of the gray whale's blubber, the migratory route of its Eastern North Pacific stock, the species' nearshore feeding habits, even its relative ugliness—all these meet the relatively modest requirements of what Latour (2014) calls "sources of an action," as they are necessary conditions without which the Makah whale hunt and the politics around it would not have taken the shape they have.[11]

Yet somehow this does not seem satisfying. There is something arresting about gray whales' apparently species-level penchant to approach boats, precisely because it seems to signal intentionality of some sort. The willingness to suffer human contact, let alone to seek it out, marks an animal as *tame*. We expect this behavior from our pets—demand it, even—but when a nondomesticated creature weighing upward of 25 metric tons and measuring over 40 feet in length does the same, it usually does not fail to capture the imagination.

Diving deeper into human enactments of gray whales has not led us closer to gray whale intentionality, however. We may ultimately have to leave that question to those who take a more direct interest in cetacean *umwelten*. What this narrative can show us are the spaces where gray whales are provided more or less opportunity to act in ways that have an impact on the stories we tell about them. We should be able to distinguish, for instance, between Makah and activist recognitions of the whale's approach, on the one hand, and the state's refusal to recognize it at all, on the other. While all three arguably represent forms of silencing in that they each attempt to speak *for* whales rather than with them, only one refuses to acknowledge the possibility—or, at least, the significance—of what whales might be attempting to communicate in those moments. There is some irony as well to the fact that the secular ontology that is most agnostic about the whale's interior life (i.e., that of the state) is the one that gives whales the least expressive liberty.

Just as one version of Makah spirituality might seem to lock prey animals into a position where they can do nothing but sacrifice themselves, the secular antihunting mode of relating to animals seems to lock human-animal relations into a state of demystified amity. Excluded from this normative picture of the world is the buzzing excitement of the chase that hunters describe whenever they pause to wax philosophical (e.g., Heavey 2013; and also clearly present in Theron's account of the rogue hunt). This thought occurred to me when I visited Jonathan Paul and Tami Drake's home in the mountains of southern Oregon and woke in the morning to find their unfenced outer backyard filled with white-tailed deer. I crept out on the porch to catch a glimpse of the animals, expecting them to turn and run at any moment. Instead, they milled about the area for hours as we spoke to them and to each other. The closest of the deer were no more than twenty feet away, on the other side of a tall wire fence meant to protect the vegetable gardens from their voracious herbivory. How different this was from what I had been accustomed to seeing while growing up as a hunter in Michigan, let alone in Neah Bay, where hunting season is essentially open all year. Staring at the deer in Jonathan and Tami's backyard, able to closely examine behaviors I had previously only glimpsed from a distance, I wondered if something might be lost in the depth of that familiarity. Without the pounding pulse of the chase, would a vegan utopia lack some elemental form of interspecies flourishing? At the same time, however, an idea came to mind that seemed to fully express the activist ontology in opposition to the Makah point of view: if, as Makah canoe builder Lance Wilkie once told me, the whale approaches because the Makah pray for it to do so, the activist retort might be that whales only ever flee because we chase them.

CONCLUSION

Antiwhaling activists must have recognized the whale's behavior in the rogue hunt.[1] It was just five years earlier that Brenda Peterson and Linda Hogan had described the scene of the first Makah hunt as "all so familiar" with "the curious whale . . . rolling sideways to gaze up at the excited people . . . so intimate, trusting" (2002, 147). The gray whale's curiosity and its well-known penchant for "approaching people" led Will Anderson to depict that hunt as a betrayal of the whale's trust in humans.[2] Yet in the weeks that followed the rogue hunt, veteran activists like Anderson mostly ignored this aspect of the whale's behavior in their public comments. Instead, the activists chose to focus on the technical breaches the hunt represented, including the violation of the Makah Tribe's memorandum of agreement with the US federal government and the tribe's seeming disrespect for its own rule of law. After Cascadia Research Collective had identified the whale as CR-175 (a putative resident/Pacific Coast Feeding Group whale with a long sighting history in the area), Margaret Owens made sure to emphasize, when contacted for comment by the *Peninsula Daily News*, that the whale was part of a "small, unique group," the existence of which, she said, the Makah Tribe had previously refused to acknowledge.[3] In the forty-three pages of comments that she submitted during the public comment period for the draft EIS the following year, Margaret did not once mention the whale's friendliness or its penchant for approaching boats.

With this rhetorical shift, Margaret and other activists tacitly acknowledged something that important actors on both sides of the conflict had come to accept: the future of the Makah whaling conflict would be decided within a moral economy in which the meaning of the whale's approach mattered little, and also where arguments about the inherent wrongness of killing gray whales could find no expression. Driven by a desire to "speak for the whales," the activists had learned to speak in a technical register that would place their concerns within the scope of the National Marine Fisheries Service's mission. They had learned to see whales like a state.

The Makah Tribal Council's disapproving reactions to the rogue hunt can be viewed as part of a similar story of adaptation and concession. It was not as if Makah officials doubted that the gray whale population could support another hunt. They trusted the advice of their marine mammal biologist, the observations of Makah fishermen, and the rulings of the International Whaling Commission, all of which supported a greater number of "removals" than the five whales for which the tribe had applied. The tribal government had based its original petition to remove the gray whale from the Endangered Species List on the unofficial observations of its commercial fishermen, who had seen what they believed to be unmistakable signs of species recovery. Nor did these officials think that the Treaty of Neah Bay should be subject to the limitations of the Marine Mammal Protection Act. But the officials had seen the gains that could be made by working within the federal fisheries bureaucracy, and their decisions on how to proceed were informed by the significant victories that treaty tribes had won in the second half of the twentieth century. Federal legitimacy for whaling was part of their vision. If other voices within the tribe had any inclination to test the state's resolve in exercising its epistemological prerogative, they were silenced by the US Coast Guard's swift and forceful response to the rogue hunt.

For almost as long as the Makah whaling conflict has been in the public eye, people and institutions on both sides have sought to legitimize their objectives by translating them into the language and logics of fisheries management. Out of a sense of practical necessity, they have adapted their arguments into forms acceptable to fisheries managers and persuasive to administrative law judges. In no way does this make the actors in this story passive victims of the state, however. Makah officials and antiwhaling activists have each effectively utilized the discourse of the state in pursuit of their respective goals. The tribe successfully staged the first whale hunt to take place in the contiguous United States in over fifty years, and they did so in a country and in a region marked by broad and diffuse affection for cetaceans. Antiwhaling activists have indefinitely delayed the hunt using a system not designed to accommodate preservationist arguments, and they helped put the Pacific Coast Feeding Group of gray whales on the map (literally), causing NMFS to rethink its concept of "stock" in the process. In a broader political context with predominantly antiwhaling sympathies and an institutional context that lacks a means to express those sympathies, the

Makah Tribe's ability to keep the possibility of whaling alive and the activists' ability to temporarily stop the hunt seem equally impressive.

These efforts at translation have not been without risk or costs to either side. Activists have found their agendas transformed by their decision to adapt to the ontology of the state, as they have moved toward arguing for the conservation value of subpopulations and stocks of whales rather than against killing whales per se. Makahs face the possibility of the state ruling against the hunt, transforming it from a sovereign exercise of treaty rights into a bureaucratically administered practice that can be perpetually delayed or, worse, ruled illegal.

"For anything less than completely revolutionary ends," goes a well-known passage from the anthropologist James Scott, "the terrain of dominant discourse is the only plausible arena of struggle" (1990, 103). Scott downplays the transformative potential of such situations on minority views because of what he perceives as the malleability of hegemonic discourses; adopting the terms and concepts of the dominant discourse, Scott argues, allows revolutionary thinkers to outwardly disavow their most threatening ideas while continuing to nurture them in private. This book has explored the insidious effects that this sort of adaptation can have, however, when the terms of struggle shift to the tacit moral economy of the state. Some ideas simply cannot be expressed in the dominant discourse, and the attempt to do so can entrench the ontological assumptions and values held by the dominant interest.

The state's interpretive framework has a rigidity that pushes back against attempts to contest or transform the meanings upon which it depends. Contemporary anthropologists tend to act as if all forms of order can be freely co-opted—as if all that is *made* can be promiscuously remade. Yet there are limits that the actors in this story have not been able to breach, things that no strategic maneuvering or contestation of meanings can change given the current legislation and the manner in which it is implemented.[4] Antiwhaling activists have been able to convince NMFS to rethink what counts as a stock, but they have not been able to move the agency one iota from its commitment to the concept of stock as the proper unit of conservation and fundamental basis for its decision-making.[5] Makah officials successfully petitioned for the delisting of the Eastern North Pacific gray whale in the space of just a few years, but the state has shown no willingness to hasten or abate its process now that delays stemming from uncertainty over a tiny, putative subpopulation of gray whales are nearing their twentieth year.

This is where the story of the Makah whaling conflict has something important to contribute to an anthropology that seeks to take itself "beyond the human" (Kohn 2013), a contribution that may be best understood within the metaphorical language favored by the proponents of that same intellectual movement. By using the same metaphor, speech, to refer to scientific as well as political representation, theorists like Bruno Latour and Michel Callon have attempted to overcome what they see as the culturally specific boundary separating nature from politics and nonhuman from human. Latour is especially concerned with "how we can go about getting those in whose name we speak to speak for themselves" (2004, 70). Such an agenda would prompt little argument from the antiwhaling activists whose informal motto is to "give the whales a voice" in the political conversations that affect them. Speaking, being heard, having a voice: these are the metaphors Americans live by when discussing political representation.

What I especially want to point out is that in this specialized form, as well as in common usage, *having a voice* is generally treated as an end in itself, the antipode to which is *silence*. "To make people talk and, if they do not do so on their own, to free their tongues and *release them from silence*" (my emphasis) is how Michel Callon describes the goal of "well-intentioned policies and the social sciences" (2010, 88). But when it comes to the political representation of nonhumans, there are outcomes that lie between speech and silence. It is not enough to say that whales always enter human social worlds accompanied by human spokespersons (Blok 2007, 84), or to ask if they may hold their representatives accountable (Kirksey and Helmreich 2010, 555), although these are important questions. Representation is not only a question of having a voice and of who speaks for whom, but in what language, literally and figuratively, one is compelled to speak.

The gray whales in the Makah whaling conflict are not silent, at least not in the sense that is generally intended by scholars of posthumanism. They have their spokespersons, those who would represent them politically and scientifically. Gray whales speak on the pages of the *Federal Register*, in stock assessment reports, in the proceedings of the IWC Scientific Committee, and in the study that has delayed the Makah hunt indefinitely. Their choice to forage in the Strait of Juan de Fuca or migrate to the Bering Sea speaks as loudly as any human voice in conversations on Makah whaling that are dominated by regulatory science. We hear them when they approach boats. When they evade the harpoon. When they die. The problem, for activists, is not that whales lack a political voice but that there is nothing that whales can say in the authorized dialect of fisheries management that does not presume the moral acceptability of their own violent deaths. One way to summarize the arc of recent history in the Makah whaling conflict is to say that antiwhaling activists learned to speak on behalf of whales in a technical language that presumes the moral acceptability of whaling in the long run. This is why the burden of translation does not weigh evenly on both sides of the Makah whaling conflict.

In the introduction to this book, I called the Makah whaling conflict a tangled mass of institutions and agendas, of ready-made narratives and interpretive frames, a place where the whale's intentions, capabilities, and materiality were debated and contested long before a single, concrete living being ever entered the story. My goal has not been to disentangle these narratives—the parties to the conflict have struggled with this enough, usually creating more tangles for themselves in the process. My goal, rather, has been to examine the nature of these entanglements, to situate them in time, and to show why some of the knots tie together more neatly than others.

In each case, Makah and activist, the threads of this narrative are conjoined to preexisting histories of struggle. For Makahs, the meaning of whaling is informed by the battles for fishing rights and tribal resource sovereignty that still simmer nearly four and a half decades after the Boldt decision (the controversial judgment that awarded half of all commercial salmon takes to tribal fishers). Those battles have taken place alongside the formation of Pan-Indian identities on the Northwest Coast, the renaissance of traditional forms of Makah spirituality, and the emergence of deliberately "cultural" forms of belonging. For the opponents of Makah whaling, the Makah hunt represents an interruption and a threat to decades of progress in the struggle to protect whales from lethal relationships of use—a struggle that has been fueled by a diffuse, growing affection for whales among the populations of Europe and the United States and often centered on the US Pacific Northwest.[6]

While by no means ignorant of one another, the actors who drive these narratives are much more interested in how their concerns have been heard and understood by the state than by each other. We must consider the role of the state in these narratives in order to transcend the overly simplistic and reductive accounts of the Makah whaling conflict that have taken hold in the press and popular imagination: to wit, that the revival of whaling represented a "return" to a traditional practice and a "reaffirmation" of Makah identity and that antiwhaling activists were moved to act by racist impulses or by misguided ideas about cultural authenticity. Both of these explanations mistake rhetoric for substance and thus fail to grasp what is at issue and what is at stake in the Makah whaling conflict.

The idea of Makah whaling as a *reaffirmation* of tribal identity and sovereignty cannot account for why some Makahs initially opposed the hunt only to later support it, notably after outside resistance was met. It cannot explain how what was once a class-based privilege of elites came to stand at least temporarily for what it means to be Makah. In chapter 1, I observed that contemporary Makah whaling—especially the persistent efforts by the tribal government and its constituents to revive whaling in the 1990s—can be viewed in a recent historical context that is complex and contingent. The initial impetus came in the wake of the cultural revival catalyzed by the Ozette archaeological discovery and the broader identity politics movement, but the actual whaling initiative was taken up by young commercial fishermen and urged on by frustrated elders. Within this context, the idea to revive whaling began as another step in the recovery of tribal resource rights, a vision of growing tribal self-determination imagined by fishermen who "wanted [their] marine mammals back." Today, even some of the strongest backers of whaling will admit that initially more than half of the tribe was opposed to it. Many saw the proposal as nothing more than an excuse for tribal council members to take lavish trips to Europe and East Asia at the tribe's expense. But once outside resistance was met, the continuation of the hunt became a form of political address in itself to those who would oppose Makah treaty rights.

In light of this narrative, it is especially discomfiting that the federal judge presiding over the trial of Makah whalers would reject the plaintiffs' defense of religious freedom on the grounds that the rogue hunt was a "protest hunt."[7] It is unclear why the judge would consider religious freedom and protest incompatible in any case, but the comment also fails to acknowledge the extent to which the entire project of contemporary Makah whaling can be viewed as a form of protest. Only the deep political significance of the struggle can account for the tribe's tenaciousness, the years of intense effort, and the vast material expenditure dedicated to the struggle for whaling rights with little promise of financial reward. People in the community felt the tribal government could not acquiesce in the face of a very familiar kind of external pressure not because whaling is inextricably linked to a timeless and unchanging Makah identity, but because giving in to external threats from the non-Native majority would mark them as passive dominated subjects—in short, as everything the past two generations of Native Americans in the Pacific Northwest had fought passionately to define themselves against.

Different memories of participation in the "fish wars" may account for some of the differences in how Makahs think about whaling in general and the rogue hunt in particular. People like Jerry Lucas would have supported whaling had the tribe just gone and done it without asking the first time. His preferred model of protest was that of an era defined by the activism of Billy Frank Jr. and the "fish-ins" of the 1960s: go in defiance of state regulations—even without tribal approval, if necessary—and fight the court battle once the rest of the community has rallied around the defendant. Others hold out hope for working within the system.

One could map this difference between the two modes of protest—between the just-go types and the work-within-the-system types—to a disagreement among contemporary Native American scholars over the most effective form of modern protest. Dale Turner (2006), for instance, has advocated the development of an intellectual community of indigenous "word warriors" who can effectively engage with the legal and political discourses of the state in order to interject Native American perspectives into postcolonial policy making. Glen Coulthard (2014), on the other hand, doubts the feasibility of such tactics. Through a reading of Hegel and Fanon, Coulthard argues that Turner underestimates the power of the state to dictate the terms in which such conversations take place, which will inevitably effect the assimilation of indigenous thought into majority ways of thinking (45–47).

The discourse on treaty rights that is central to Makah whaling may at first seem to belong to the tool kit of the word warrior. A treaty signed by a desperate and disease-worn community, written in English by a colonial power with only their promises as guarantees, and disregarded almost immediately upon its signing might at first sound like a deeply compromised and fanciful document, the precursor to what Coulthard critiques as the unrealistic aspirations of today's word warriors. But there is a sincere conviction among Makahs today that the Treaty of Neah Bay was a triumph of forward-thinking ancestors who were able to set the terms of the negotiation despite their sharply disadvantaged position. This is why the fabled canoe ride shared by the Makah chief Tse-kauwtl and territorial governor Isaac Stevens during treaty negotiations is so significant in today's retellings. What might have been destined to be a historical footnote, the canoe ride moves the treaty negotiation in today's memory into a distinctly Makah space, a perfect setting for the whaler-chief's uniquely Makah request: "I want the sea, the sea is my country."

The struggle for whaling rights today can be seen as a special type of what I call, in chapter 2, the politics of *yum*: the curious fact that the very articulation of "Makah" attitudes toward whales can have a purposefully provocative effect on non-Native listeners (or so it is believed). This *yum* response, as I call it for short, reflects a crucial moral difference between Makah attitudes toward whales and those associated with what we might call the American mainstream. Whales have become, as Anders Blok (2007, 66) calls them, "near-sacrosanct agents of empathy and protection" for most Americans. In Neah Bay, meanwhile, even the local opponents of whaling tend to view whales as things-to-be-eaten, reflecting the utilitarian assumptions toward animals that are widely shared in the community.

These utilitarian attitudes are part of the reason I reject the claim that whales are properly considered "nonhuman persons" for Makah whalers, a claim that is widespread and influential enough to deserve additional comment here.[8] In my view, the use of the term *nonhuman person* is deeply mistaken and risks straying into what Mary Weismantel calls "a retrogressive return to romantic primitivism that sometimes mars the ontological turn" (2015, 142). As Carlos Londoño-Sulkin (2012) has pointed out in reference to the relational ontology of Amerindian hunters, personhood implies a moral equivalence with human beings that is simply not present in the thinking of the Upper Amazonian group with whom Londoño-Sulkin works, for whom comparisons between humans and animals are often made in a pejorative sense at the expense of the animal (and of the person being compared to the animal). Hence the use of local terms that roughly translate to "real person" to describe the human occupants of the social group. It is telling that Inupiat, the autonym for another North American whaling people that have been described as animists, also literally means "real people."

To argue that Makah whalers think of whales as persons makes sense only if one jettisons the moral denotation of the English word. I have been at pains to show that this moral dimension is precisely what matters if we want to understand what is at issue and what is at stake in the Makah whaling conflict, particularly with regard to how the conflict has been understood and managed by the state. The moral dimension of personhood is crucial for understanding how some Makah whalers view whales in spiritual terms and as commodities—an enactment of sacrality that also perhaps best goes by another name—whereas Euro-Americans, for whom sacredness means "set apart," tend to see this as a contradiction (Nadasdy 2005). For the latter, but not the former, as Kopytoff observes, "the conceptual distinction between the universe of people and the universe of objects [has] become culturally axiomatic" (1986, 84).[9]

If not persons, then what are whales within Makah spiritual discourses? In chapter 3, I traced the multiple ways in which Makah whalers make sense of whales as they become entangled in spiritual narratives. I observed that variability and secrecy are the rule rather than the exception for Makah spirituality, and that spiritual practice can push back upon the present, providing the kind of "skills" that Timothy Ingold (2000) identifies as integral to the environmental imagination of hunters. Spiritual preparation was a significant part of the context within which harpooner Theron Parker made sense of his experience with whales.

A vibrant local discourse suggests that Makah administrators were well prepared by their traditional and spiritual teachings to inhabit the role of fisheries manager, which partly explains why it never was the intention of the tribe to reject the jurisdiction of the state altogether. In chapter 4, I explored how the good working relationship between Makah leaders and federal fisheries managers runs against default anthropological assumptions about the compatibility of so-called traditional ecological knowledge and modern resource management. Makah leaders find in the state a source of political power through which they can accumulate resource rights while translating, preserving, and instrumentalizing traditional understandings of the nonhuman world. Because of the moral consonance between the two ways of seeing whales—each viewing whales, in the most abstract sense, as killable resources—Makah leaders have been able to pursue federal legitimacy in the form of resource co-management without disrupting or undermining the spiritual and ontological commitments of Makah whalers.

For activists, a different kind of connection with arguably a different kind of whale drives their participation in the Makah whaling conflict. Chuck Owens spoke of the "majesty" of whales, and Margaret Owens challenges us to think of what other animal captivates people with its very breath. Chuck and Margaret were teenagers when attitudes toward whales changed dramatically across much of the Western world, as whales went from being little more than living containers of commercially important oils to beloved icons of the environmental movement in the space of little more than a decade. It was "in the air," Margaret says of the growing, global affection for whales and dolphins during her teenage years. Translating these feelings into action has been especially challenging for activists. How does one convince others not to harm something because you find it beautiful or magnificent? This would be difficult in any case, but it is nearly impossible within the moral economy of NMFS. Indeed, one of the most striking things we can learn from the Makah whaling conflict is that for the state, the question of whether humans ought to be able to kill whales is already decided in advance. For the administrators at NMFS whose task it is to oversee the recovery and management of whale populations, gray whales fall under a sweepingly broad system for managing commercially important ocean resources.

In spite of shifting public sentiments toward whales and dolphins over the past fifty years, the US federal government continues to manage whales essentially as if they were large fish. Still, as attempts to directly intervene in the hunts proved more and more untenable due to the US Coast Guard's seizure of activist vessels in the 1999 and 2000 hunts, activists perceived the need to adapt. The evolution of their tactics provides a succinct demonstration of the workings of technocratic authority on grassroots animal activism in the early twenty-first century. Given the crackdown on animal activism in US federal legislation since that time, it is likely to be an enduring portrait.[10]

Despite these tensions, it is easy to see why people on both sides see promise in pursuing their goals within the medium of fisheries management. Activists see in regulatory science the paradoxical promise of perpetual, temporary victories underwritten by irreducible scientific uncertainty. In this pursuit, they join a long history of actors on both sides of whaling debates in exploiting the "unclosable loophole" of scientific uncertainty.[11] Makahs see in fisheries management a discursive space where it is possible to win federal approval to hunt whales and gain the protection of a military arm of the federal government, all despite the prevailing preservationism of American attitudes toward whales.

The Uneven Binding of Recognition

Another way to frame the argument of this book is to express it in terms of the relative benefits of *recognition* for Makah whalers, antiwhaling activists, and whales. Is it possible that Makahs derive advantages from the formal recognition of the state, or have they fallen victim to what Glen Coulthard (2014, 48) calls the assimilative lure of the statist politics of recognition? Considering the Makah whaling conflict in these terms helps to identify some of the more general factors that ultimately decide who benefits from recognition.

Critical theorists in the field of indigenous studies have tended to doubt whether indigenous groups operating within what Simpson calls "the teeth of empire" (2014, 158) can ever truly benefit from the state's formal recognition when they remain in the thick of a violent, asymmetrical, and ongoing colonial encounter. If the overriding purpose of Indian policy remains the dispossession of lands and resources and the assimilation of indigenous people (as Coulthard suggests is true of the Canadian federal government's handling of aboriginal groups), then Native peoples' acceptance of the terms of struggle could mean accepting the conditions of their own eradication (2014, 25, 42; see also Simpson 2014, 157–58). Yet sometimes indigenous groups *do* benefit from formal recognition, and since it is reasonable to doubt that these are random occurrences, it follows that recognition is not *necessarily* antagonistic to the interests of indigenous groups or persons.[12] We can expect the ramifications of recognition to depend on the terms and conditions upon which it is granted and the moral economies at play.

One important criticism of the politics of recognition concerns its tendency to freeze legitimate indigenous practice in a precolonial past, especially when culture rather than political independence or sovereignty serves as the basis for certain rights or exemptions (Simpson 2014, 20). It makes matters worse when the concept of culture in question is the totalizing, early twentieth-century kind favored by institutions like the IWC—although, as Markell (2003, 160) has pointed out, such contradictions and double binds will always be the end result of a politics of recognition as long as culture remains the basis of rights claims not because of anything unique to the concept of culture but because of the temporal (backward-looking) character of *identity*, which is immanent in action and only ever coalesces in retrospect (13–15). "Who we are" has no other basis than "who we have been," which, when formalized by the state as the basis of rights, becomes a discourse on authenticity. Responding to public criticism of the tribal whaling crew's use of modern implements in the 1999 whale hunt, Janine Bowechop reflected on the apparent hypocrisy of the situation:

> A lot of people just like to freeze us in the past . . . but we allow other cultures to make changes. One of my friends says, "I'm a, y'know, white American but I don't make my butter in a butter churn anymore and I'm not criticized for that." Folks don't ride in covered wagons anymore and we don't turn around and say "Gee, you're not a real American." But, unfortunately, we're continually criticized if we do anything different than we did five hundred years ago.[13]

Along with this expectation that cultural reclamation will "look traditional" (Deur 2000), another common aspect of authenticity discourse is the assumption that culturally meaningful activities will enjoy the full support of a community. This is a corollary of the implicit view that indigenous politics are consensual rather than dialectical, and it makes the appearance of internal dissent look like inauthenticity.[14] Moreover, any attempt to jettison an essentialist concept of culture and adopt a more dynamic understanding of the concept, as Bowechop and the Makah Cultural and Research Center have tried to do (see also Coté 2010, 153), cuts the legs out from under the original claim. If culture changes, then what is the basis for preserving or "bringing back" lost traditions? What is the stable subject position from which the claim to cultural exemption is made?[15] Attempts to evade the petrifying politics of authenticity only leave the claimants asserting rights on a foundation that is still essentialist but ever more ethereal and abstract, a true double bind.[16]

The situations that produce this sort of authenticity discourse share two additional features, however, without which there may be no double bind. First, they involve an exemption to a rule of general applicability (otherwise there would be no conflict); second, they require the party seeking the exemption to formally establish some kind of authenticity, whether of an identity or practice (e.g., a particular fishing method or subsistence pattern) to the satisfaction of some outside arbiter.[17] This was a feature of the Makah Tribe's application to the IWC that is absent from the domestic reauthorization process. Federal approval is not tied to questions of authenticity. It is crucial to note that the criticism to which Bowechop is responding, above, and which has understandably attracted a great deal of attention from scholars for its troglodytic character[18], tends to show up exclusively in places that may be irritating to tribal leaders but are ultimately nonthreatening, legally speaking: scrawled in the letters to the editor of local newspapers, screamed at public protests, or muttered to journalists. No one in any position of authority within the National Marine Fisheries Service has an incentive to take such comments seriously in the execution of their official duties, and for at least the last thirty-five years, no one has. Far from the sovereign pronouncements of a powerful white majority, then, these comments would seem to issue from a place of powerlessness (although that does not excuse their ignorance).

An alternative reason to suspect that formal recognition would undercut rather than favor the reauthorization of Makah whaling is the assumption that the laws of the settler state are always predisposed to undermine the interests of indigenous peoples. Audra Simpson argues, for instance, that recognition and inclusion are only ever extended by the settler state "*if* the problem of cultural difference and alterity does not pose too appalling a challenge to norms of the settler society, norms that are revealed largely through the law" (2014, 20). Setting aside the implication that law necessarily reflects prevailing norms, the application of this expectation in this case would seem to have it backward. It is the protection of nonhumans that offends the current state of the law and its default imperative to encourage the circulation and commodification of nonhuman nature.[19] If the law enshrines any moral disposition toward animals, it is that of *use* over nonuse, of conservation over preservation.

This case is slightly complicated by the fact that the Makah Tribe is not actually requesting an exemption to a law of general applicability but an *exemption to an exemption*. The Marine Mammal Protection Act is something of a vexed document because of its vestigial preservationism, and regulators have struggled to implement those precepts using the conservationist framework of fisheries management, which favors maximum sustainable use. As if it were a double negative, the Makah request aligns the tribe with the default value system of the state—with an ethic of maximum use.[20]

Whales, too, operate within the teeth of empire, within what Derrida (1991) calls a sacrificial structure that relegates all nonhumans to the category of killable beings. Antiwhaling activists' attempts to represent gray whales, to speak for them, to win them recognition in this system actually bind them into what is, from the perspective of activists, a hostile framework, putting into motion exactly the sort of scenario that Coulthard feared for indigenous actors (2014, 78). Entangled in what he calls "the power-laden process of discursive translation," meanings become reoriented in ways that entrench the moral assumptions of the dominant actor, reinforcing instead of undermining state sovereignty (see also Markell 2003, 30). Whether those values serve the interests of those it binds, however, varies depending on the factors I have described here.

Lost in Translation

At the most recent public comment sessions hosted by NMFS in April 2015 in Seattle and in Port Angeles, concerned citizens settled back into their well-worn habits of talking past one another. The regional administrator for NOAA Fisheries West Coast started the presentation by saying "a bit about the whales themselves." After introducing the recent history of the tribe's request and providing a brief rundown of the relevant federal legislation, she mentioned the gray whale's conservation status (neither endangered nor threatened), its migratory habits (from Baja to the Bering Sea, but sometimes stopping in Washington), and the total number of animals in the "herd" (around twenty thousand) and in the Pacific Coast Feeding Group (around two hundred). A brief period for public comments followed. Fourteen people signed up to make a comment at the Seattle meeting; of them, four "unaffiliated individuals" spoke out against the hunt on the grounds that the whales had become unafraid of boats. These four individuals talked of the level of comfort the whales had reached with humans that made them "vulnerable," the whale's changing behavior after encounters at the birthing lagoons in Baja California, and the expectation of friendly human contact that would make the whales "fish in a barrel" and the killing of them a "breach of their trust."

"All this is for nothing," said a commenter who introduced herself as a Chippewa from North Dakota. "[The Makah] have already won. They have it in their treaty and their treaty comes before your Constitution."

Three activists with longstanding involvement in the conflict spoke at the two events. The executive director of Sea Shepherd Legal expressed skepticism over the legitimacy of the IWC's approval for the hunt and mentioned other conservation issues that were not sufficiently addressed in the environmental impact statement. She mentioned the unresolved scientific uncertainty around gray whale populations and the risk to the endangered Western North Pacific gray whale when "it is virtually impossible to tell them apart."

Will Anderson asked rhetorically how many of the Pacific Coast Feeding Group would have been killed had the Makah hunt been allowed to go forward as originally planned. He suggested NMFS take into consideration the previously unclassified "North Puget Sound population" of gray whales, currently numbering around eight. Finally, unable to help himself, he proposed that the word "harvest" be replaced in the draft environmental impact statement (DEIS) with "harpooning and shooting a gray whale."

"This is a scientific document," he said. "Let's call it what it is."

Only one commenter, an unaffiliated individual, number five of fourteen at the Seattle meeting, had the temerity, or perhaps the naivety, to directly challenge the agency's assumptions about whales. "I don't know how you can take members of our community that are not human but are part of the human environment," she said, "and [treat] them [as] a resource or a commodity, like lumber in a forest."

"Thank you," the moderator replied dryly before calling on the next person.

In 2015, Steve Stone, an official from the NMFS West Coast Protected Resources Division, published the agency's responses to a selection of public comments on the 2008 DEIS. The DEIS elicited over eight hundred pages of written comments from more than four hundred different individuals and organizations, according to the document, with topics ranging from "specific biological, ecological, or legal issues to more general cultural, personal, and spiritual values." Generality and specificity sound like euphemism here; and indeed, the latter more "general" comments—which appear to include all moral objections to whaling per se—made up a substantial portion of the total comments received by the agency but exactly zero percent of the comments to which NMFS selected and gave a substantive reply.[21] "In some cases," read a sentence in the preamble, "topics were outside the scope of the DEIS."

A truly benign way of putting things, this statement should surprise no one who reads it. What it does not acknowledge—and why we might suspect the agency of false modesty in this instance—is that from the perspective of whalers and activists, there is no "beyond" that is outside of the DEIS, no alternative forum in which their concerns might be heard. People on both sides remember what happened when they attempted to take matters into their own hands, how swift and violent was the response of the state. More than a delimitation of bureaucratic responsibility, the statement reflects a narrowing of vision with profound moral consequences, excluding from consideration the bulk of what Makahs and antiwhaling activists think and feel about gray whales. Outside the scope of the DEIS, banished into bureaucratic limbo, are the enactments of the whale as a moral subject, as a spiritual being, as anything other than a natural resource or statistic. Definitively excluded from consideration are any means of ascertaining the wisdom of a hunt outside of secular epistemologies as well as the activists' arguments for protecting whales regardless of the health of their "stock." There is nothing unique about the US federal government's disposition toward gray whales that determines this outcome; on the contrary, one of the most striking things the Makah whaling conflict shows us is that whales are not special. The state's rigid ontology applies even to the most iconic of animals, the most charismatic of megafauna, and the most meaningful of Makah traditions.

ACKNOWLEDGMENTS

The practice of ethnography depends on the generosity of others. I am deeply grateful to those who shared their stories, their opinions, and their knowledge with me. From the Makah and Neah Bay community, I especially want to thank Vicky and Dan Druge, my hosts for several visits, for offering me a home away from home. I also want to thank Matthew Whitacre for his friendship and hospitality. Among the informants for this project, I am especially grateful to Greg Colfax for his time and his advice. I do not know what life in Neah Bay would be like without him there, pondering his latest beautiful piece of artwork, puffing on his pipe, and sharing his kindness and wisdom with friends and passersby. Micah McCarty and his wife Haley were both exceedingly generous with their time as well. I do not think anyone has ever spoken at length with either of them without feeling as if they were changed in some way for the better. Wayne Johnson helped me at various stages of this project and has never been anything but kind and generous. I consider him a friend.

This project would not have been possible without the magnanimity and remarkable memories of Margaret and Chuck Owens, whose passion and knowledge on this issue are unrivaled. I would also like to thank Will Anderson, Jonathan Paul, and Tami Drake. Jean Vitalis is a leader in her community, and I can understand why—her intelligence and honesty made for one of my favorite interviews. I found Chad Bowechop to be among the most open and reasonable people I've ever spoken with, and Dave Sones earned my deep respect with his articulate candor. I would also like to thank Keith Johnson, Theron Parker, Andy Noel, Donnie Swan, Arnie Hunter, the late Ben Johnson, Darrell Markishtum, the late Gary Ray, Jerry Lucas, Joe McGimpsey, and Greig Arnold, as well as others whom I cannot name for reasons of confidentiality.

I was very fortunate to encounter other scholars during my time in Neah Bay whose advice and contributions made my life and this project easier. Jennifer Sepez continually offered valuable advice that only she could give. Kim Wilson is an excellent historian and an astute observer of social life. Thank you also to Ylva Hernlund for her timely advice. Ann Renker graciously agreed to meet with me several times during my fieldwork and provided me with a number of helpful suggestions.

It was a pleasure to complete this book at the Williams College Center for Environmental Studies, where Nick Howe, Pia Kohler, and Ralph Bradburd offered me invaluable professional and academic advice, as did Denise Buell and Rhon Manigault-Bryant. I will forever appreciate the welcome and support I received from the Anthroplogy and Sociology Department at Williams—Joel Lee, Antonia Foias, Olga Shevchenko, Grant Shofstall, Christina Simko, David Edwards, Lisa Koryushkina, Jim Nolan, and Peter Just. I am especially grateful for the support of the Oakley Center for Humanities and Social Sciences, Jana Sawicki, and Krista Birch for hosting a review of this manuscript and bringing in the wonderful Harriet Ritvo and Michael Rossi to participate. I wish every scholar could have such an opportunity, and the Oakley Center does an excellent job of making this opportunity broadly available to the faculty at Williams. Thanks also to Megan Grabill for her contribution as a research assistant at Oberlin College.

No academic experience may ever surpass the year I spent at the American Academy of Arts and Sciences, where this project was transformed by the sage advice of Larry Buell and fellows Lukas Rieppel, Rachel Wise, Katherine Marino, Merve Emre, Joy Rankin, Rachel Guberman, and Emily Owens—all part of the visiting scholars program capably run by John Tessitore. Reflecting on my time there also affords me the opportunity to acknowledge the brilliant Harriet Ritvo a second time. While in Cambridge, I had the privilege of participating in Sheila Jasanoff's Science, Technology, and Society Fellows program at the Harvard Kennedy School as well, where I benefited from the feedback of Zoe Nyssa, Sam Weiss-Evans, Margo Boenig-Lipstein, Antony Adler, Zara Mimalek, Gili Vidan, Matthew Sample, and many others.

This project began at the University of Chicago, and it would not exist in its present form without the U of C's unequaled commitment to intellectual rigor and academic freedom. Rick Shweder deserves my greatest thanks. His brilliance, collegiality, and commitment to the examined life were matched by his unwavering support for me and my work. No one contributed more to this project than Don Kulick. I cherished every conversation I had with him. Thanks also to Richard Taub, Mark Payne, Jennifer Cole, John Lucy, the late Bert Cohler, Eugene Raikhel, and the rest of the Comparative Human Development faculty.

My sincere thanks to all of my friends and fellow graduate students at the University of Chicago whose inspiring conversation provided nothing less than an independent reason for being; in particular among my graduate cohort: Nadxieli Toledo Bustamante, Juan Jose Bueno Holle, Julia Kowalski, and Katie Jenness. To Nadxieli in particular, I hope I have returned a fraction of the friendship and support that you have given me. Jacob Hickman has been my close friend from almost the moment I arrived in graduate school. Also, in no particular order: Hallie Kushner, Teofilo Reyes, Talia Weiner, John Davy, Amy Sousa, Eli Thorkelson, Lainie Goldwert, Roscoe Nicholson, Barnaby Riedel, Christine El Ouardani, Carly Schuster, and Mike Kaufman. I would like to thank the Bueno family for their hospitality during my long drives to and from my field site, as well as my friend Jason Glover.

The Animal Studies Workshop at the University of Chicago showed me what interdisciplinary discussion and collaboration can look like in their ideal form. Heather Keenleyside, the original faculty sponsor of the group, is a model of what I would like to be as a scholar and a mentor. I am particularly grateful to Joela Jacobs, Katharine Mershon, Agnes Malinowska, Bill Hutchison, Sam Schulte, and Jessica Robinson for their feedback on my work.

I would be remiss if I did not mention the speech and debate program at Central Michigan University, where I was introduced to the concept of an academic argument. Ed Hinck, the program's longtime director, provided me with indispensable career advice on numerous occasions. He has my highest thanks, as does my former coach, John Boyer, and all of my teammates.

There is nothing I can say about Ashley Drake that would convey the true measure of her contributions to this project as my partner, friend, and confidant. To my parents, Les Sr. and Susan Beldo: all that I accomplish is a tribute to your love and support.

Previous versions of chapters and arguments in this book were presented at the meetings of the American Anthropological Association; the American Ethnological Society; the Society for Psychological Anthropology, the Harvard STS Circle; "After Extinction," a conference at the University of Wisconsin–Milwaukee Center for 21st Century Studies; "Navigating a Multispecies World: A Graduate Conference on the Species Turn" at Harvard University; the workshop in Comparative Human Development at the University of Chicago; and the Animal Studies Workshop at the University of Chicago. Special thanks to the colleagues I met at these events who provided critical feedback on the project, including Amy Field, Daniel Solomon, and Mary Weismantel.

Funding for this project was provided by the Andrew W. Mellon Foundation, the American Academy of Arts and Sciences, the Social Sciences Division at the University of Chicago, the Giannino Fund, the Mary Anne North and James L. Rynerson Research Fund, and the Department of Comparative Human Development at the University of Chicago.

NOTES

INTRODUCTION

1. The belief that prey animals will sacrifice themselves to human hunters has been widely documented among northern hunting peoples; see Brightman 1993; Tanner 1979; Nadasdy 2007; Ingold 1989; Willerslev 2007. Brightman (1993) refers to this set of beliefs as "grateful prey."
2. See Escobar 1999; Descola 2013.
3. Ingold 2000, 41.
4. Latour (1987) writes, "Since the settlement of a controversy is *the cause* of Nature's representation and not the consequence, we *can never use the outcome—Nature—to explain how and why a controversy has been settled*" (99; italics—all of them—in the original)
5. See, e.g., Escobar 1999, Butler 1990, and Mol 2002, respectively. I prefer Mol's term, *enactment*, over the more popular term, *construction*, because it brings attention to the practices through which nature is formed and sustained while remaining agnostic about the actors involved and the materials that are used. "Construction metaphors," as Mol points out, "suggest that the material is assembled, put together, and turned into an object that subsequently goes out into the world all by itself" (32).
6. I am aware that calling the whale by the name given to it by a Makah opponent of the hunt could make me appear partial, but it is not my intention to take sides. It is convenient to call this whale something, and Yabis seems appropriate. A contested name for a contested whale.
7. I borrow this last phrase from Jacques Derrida, who gives the example of a cat, his cat, the cat that stares at him one day as he exits the shower. A cat that "does not appear here to represent, like an ambassador, the immense symbolic responsibility with which our culture has always charged the feline race," but is instead an "irreplaceable living being." "Nothing," he writes, "can ever rob me of the certainty that what we have here is an existence that refuses to be conceptualized" (2008, 9). My point here is that we must observe the limits to the sense that nature even in its materiality is "always already social" (cf. Kosek 2006, 28).
8. See Kirksey and Helmreich (2010) on the concept of "becomings" in the emergent field of multispecies ethnography. As Celia Lowe writes, "The idea of *becoming* transforms ideas into events, objects into actions" (Kirksey and Helmreich 2010, 546).

9. Contemporary anthropological thinking tends to attribute the idea of multiple realities to the so-called ontological turn, but it is worth noting that Shweder (1991) introduced the idea of multiple objective worlds two decades earlier in a way that was in keeping with the culture theory of the time; this is another reason why I am tempted to agree with Candea (2010) that *ontology* is just another word for a return to a more robust concept of culture.
10. Sullivan 2000, 22.
11. Here, I borrow again from Derrida (2008), who presents his own animal ethics as a call to have the animal "heard in the singular" (2008, 47).
12. See Scott 1998, Fitzgerald 2003, and Telesca 2017 on forestry, agriculture, and fisheries, respectively.
13. Whales are perfectly good fish in many respects, of course, as the philosopher of science John Dupré (1999) has pointed out. *Fish* is more of a folk category than a scientific one, and both categories (fish and whale) are paraphyletic. Sperm whales and bottlenose dolphins share a closer common ancestor than either does with blue whales, but the category of whale customarily excludes the smaller toothed cetaceans. The category of fish must include mammals, ourselves included, if it is to capture everything we think of as fish and remain defined by a single common ancestor. However, the idea of a moral equivalence between whales and fish diverges from the current prevailing opinion in the United States; see, e.g., D'Amato and Chopra (1991) on whales' "emerging right to life." This is why I think this tacit classification should evoke some surprise.
14. See Cassaniti and Hickman 2014; Fassin 2012; Laidlaw 2001.
15. In his ethnography of Runa hunters, Eduardo Kohn (2013) refers to the "tricky business" of hunting within an "ecology of selves" without otherwise indicating that the animist hunters he worked with felt any sense of ambivalence at all over the act of hunting and killing (118). Phillipe Descola, also in a study of Amerindian (Achuar) hunters, unfavorably compares the "passing prick of conscience" experienced by Westerners to the deep metaphysical disturbance felt by animists (2013, 286). In so doing, both authors present moral sentiments as if they were the logical entailment of ontological classification.
16. See Bessire and Bond (2014) for a critique of this expectation of stability among the works within the so-called ontological turn.
17. Renker 2012.
18. The time depth of both the Makah occupation of Cape Flattery and the practice of whaling in the area is contested. The Makah language is part of the Wakashan language family, which is widely distributed across Canada's Vancouver Island but found nowhere else in the United States. Contemporary language patterns suggest that the Makah language started to diverge from other Wakashan dialects about eight hundred years ago. Any attempt to assess the relative importance or purpose of whaling prior to the late precontact period is impractical due to the inherent limitations of the available archaeological evidence; for a discussion of those limitations, see McMillan (1999, 176) and Isaac (1988).
19. On the stratification of precontact indigenous societies on the Northwest Coast, especially of those belonging to the so-called Nootkan cultural type, see Sapir 1915; Drucker 1951; Isaac 1988. For the way that whaling reflected and reinforced this stratified social order, see McMillan 1999, 139; Drucker 1951; Coté 2002; Huelsbeck 1988; Reid 2015, 10–11; Swan 1870, 11.

20. On the decline of Makah whaling following the signing of the Treaty of Neah Bay, see Renker and Gunther 1990; Gibbs 1877; and Swan 1870. On the waning strength of tribal affiliations in the early 1900s, see Harmon 1998.
21. Petersen 1999.
22. 384 F. Supp. 312 (W.D. Wash. 1974).
23. On opinion polls, see Lavigne et al. (1999, 11) and Kim (2015, 244). On the ubiquity of whale watching and the diffuse affection for cetaceans in the Pacific Northwest, see Moore (2008).
24. See Russell 2001, 106–7.
25. The IWC's concept of culture was not a historical accident. As Raibmon points out, discourses on cultural authenticity tend to cast Indians as having consensual rather than dialectical civic life, part of the dominant portrayal of Native Americans as people without history or politics (2005, 12).
26. Sullivan 2000.
27. *Metcalf v. Daley*, 214 F.3d 1135 (9th Cir. 2000).
28. Environmental impact statements are significantly longer and more rigorous than environmental assessments. The 1997 environmental assessment for Makah whaling was forty-eight pages, whereas the 2008 draft environmental impact statement was over seven hundred pages.
29. For an official report from the tribe's marine mammal biologist that covers many of these details, see Scordino 2007.
30. Cf. Wynne 2010, as well as other theories for how the modern state legitimates itself through nonviolent means: e.g., as a source of order (Markell 2003, 30) or as a biopolitical exercise in human welfare (Kosek 2006, 96–102).
31. Briefly, the difference between these calculations for fisheries and cetacean management has to do with the target population level of the managed stock. PBR is set relative to the estimated "carrying capacity" of the environment, or the maximum population of a species the environment can support without adverse effects. NMFS calculates fisheries quotas based at or near "maximum sustainable yield," which sets the "optimum sustainable population" at a fraction of the environment's carrying capacity (usually the minimum, hence "maximum" yield; see Thornton and Hebert [2015] on how maximum sustainable yield models create damaging assumptions about fish harvests that portray "uncaught" fish as waste).
32. See the NMFS stock assessment for the Eastern North Pacific gray whale published in 2015 (NMFS 2015b).
33. Tribes are classified as domestic dependent nations according to US legal precedent (*Cherokee Nation v. Georgia*, 30 US 5 Pet. 1, 1831) and are subject to a kind of legal guardianship that entails certain trust responsibilities on the part of the US federal government. On ethnic groups behaving like for-profit businesses, see Comaroff and Comaroff 2009.
34. See Swan 1870, 59.
35. The *Seattle Times* has continued to receive letters to the editor that use racial stereotypes to condemn Makah whaling (Westneat 2015), and some of the more diffuse opposition in the region might also be attributed to a general sense of resentment still smoldering from the Boldt decision, which some white residents view as the awarding of special privileges to Native fishermen.
36. For an example of a scholarly work that affirms the rhetorical fiction, see Marker (2006), especially in passages such as the following: "the *Makah chose* to return to

hunting whales because they desired a reconnection to an ancestral and mythic relationship to an animal that has sustained them both physically and spiritually" (497; my emphasis).
37. Westneat 1996.
38. As Markell (2003, 153–54) points out, there are structural reasons for why this concept of culture tends to hold the most weight in such institutional conversations, and it has to do with (1) the backward-looking nature of cultural exemptions and (2) the fact that such exemptions are viewed as apolitical exceptions to what Markell calls "the eminently political activity through which we [actually] become who we are."
39. See also D'Andrade 1995. For a contrasting view, see Scheper-Hughes 1991.
40. Gail Landsman reported an identical feeling in her experience with a very similar dispute setting, what she called the Indian-white conflict at Ganienkeh (1988, 13).
41. While I think it is true that anthropologists always stand in definite relation to their subject (Abu-Lughod 1991, 468), this does not make anthropologists incapable of forming views that reflect something other than their subject's position. The very commitment to ethnographic neutrality, as Wilson (1970) points out, is part of what makes ethnography different from learned opinion.
42. Given the emphasis on nonhuman agency in some quarters of anthropology, I was mildly shocked to read Martello's claim that gray whales have *"been* globalized" (2004, 263; my emphasis).
43. See Levy and Hollan (1998) on person-centered interviewing, which informed my approach to the ethnographic interviewing I conducted for this project.
44. Because they and their opinions on whaling are already publicly known, many of my informants requested to be identified by name in this book.
45. Erikson 2002, 51.
46. Cf. Trouillot 2003 on the "savage slot" in which anthropology long placed the Other.
47. Neah Bay is a community where "culture" is spoken of as if it had clearly defined edges, as in the more humanistic rather than anthropological sense of the term. During a tribal talent contest at the community gym, I once listened as an emcee introduced a young contestant by announcing that she had been "culturally active since age five." Buying meat from Costco is very common but decidedly not "cultural," while getting fish from a family member is.
48. See, e.g., "The Makah Whaling Tradition," on the Makah Tribe's website (http://makah.com/makah-tribal-info/whaling/).
49. At no point in this book do I intend for the word *political* to be read as excluding *cultural, spiritual, moral,* or *traditional.* I find Audra Simpson's description of her informants to be familiar and apt: "The people that I worked with care deeply about ceremony and tradition, but hinged those concerns to the language and practices of nationhood, citizenship, rights, justice, [and] proper ways of being in the world" (2014, 97–98).

CHAPTER ONE

1. On Makah whaling as a reaffirmation of identity and sovereignty, see, e.g., Coté 2010, 6.
2. According to legend, Thunderbird captured the whale during a time of great famine to feed the Makah people. For a discussion of the image and its place in Northwest Coast cosmology, see Webb 1988, 16–19.
3. Bock 1995.
4. The proportion of Makahs who at first supported the hunt is contested. Coté (2010)

reproduces the Makah Tribe's claim that "over 76 percent of tribal members voted in favor of reviving whaling practices" in a 1995 referendum (2010, 131). Peterson (1996) wrote in the *Seattle Times* at the time:

> [Alberta] Thompson and [Dotti] Chamblin [two prominent Makah elders who opposed the hunt] criticized the lack of consensus and consultation on the part of their tribal council. Pushing for whaling, "the council went ahead without the consent of the tribe," Thompson said. "They say they have 70 percent [in support of whaling], but that isn't so." According to Thompson and Chamblin, there is no Makah tribal consensus on the issue of whaling, though the tribal council has officially stated to the news media that they are acting for the whole tribe. When an impromptu vote was called, only 104 . . . Makahs cast their vote and only 70 voted yes.

My own assessment of the relative support for whaling within the Makah community is based on a range of qualitative data including direct statements from supporters as well as opponents of whaling within the tribe. I am not making a quantitative claim about the exact levels of support or opposition (or ambivalence) for whaling in the community.

5. Glen Coulthard (2014) has argued that what we might call essentialist claims ought to be viewed differently within narratives of cultural survival in settings with all-too-real histories of genocidal state violence (98). In a critique of antiessentialism as it pertains to indigenous politics of recognition, Coulthard argues that constructivist criticisms ought to give more consideration to the "deeply entrenched power relations at play" in the relationship between colonizer and colonized. He is especially disparaging of those critiques that blame strategic essentialism for furthering class and gender inequality without acknowledging the broader oppressive social relations that perpetuate such conditions, noting that "the reification and misuse of culture in this case cannot be understood without reference to the colonial context within which it continues to flourish" (93–94).

6. For examples of accounts that have done little more than stamp this sort of rhetoric with the imprimatur of the academy, see Miller (2000) and Marker (2006).

7. In a quantitative analysis of Makah public rhetoric leading up to the 1999 hunt, Barton (2000) found that "as external opposition to the hunt increased, tribal members chose to fashion a more passionate discourse on the links between whaling and treaty rights."

8. Kulick (2003) shows how simply uttering the word "no" in certain contexts has significant performative stakes for the individual. While Simpson highlights a case of an indigenous refusal as an alternative to state recognition (2014, 177), Makah leaders have been willing to accept state sovereignty in order to work within existing bureaucratic channels (see chapter 4). See Nesper's *Walleye War* (2002) for a similar situation involving the Ojibwe struggle for spearfishing rights in northern Wisconsin.

9. US officials gave the Makah Tribe its official name and in so doing created the Makah Tribe as a legal entity in 1855 during treaty negotiations. According to Colson (1953), the government interpreter for the treaty negotiations was from the neighboring Clallam community, and "Makah" was the Clallam designation for the Nootkan-speaking peoples of the villages near Cape Flattery. Makahs have long insisted that their actual autonym is Kwih-dich-chuh-ahtx, meaning "People of the Cape." Swan (1870) appears to have been among the first to reference this name in print, which he rendered as Kwe-net-che-chat (1870, 1).

10. Colson 1953; Drucker 1951; Huelsbeck 1988, 1994; Jewitt and Alsop 1815; McMillan 1999, 139; Renker and Gunther 1990.
11. Colson 1953; Drucker 1951; McMillan 1999, 159–60; Renker and Gunther 1990; Sapir 1915; Swan 1870; Reid 2015; Coté 2010, 22. I deliberately use the term *stratified* here as opposed to *ranked* following Isaac's discussion of the important differences between the two terms (1988). Isaac suspects that scholars of the Northwest Coast have used the term *ranked* in order to avoid the "thorny question of social stratification."
12. Drucker 1951; Coupland 1998, 51.
13. Reid 2015, 10; Drucker 1951, 272–73; Renker and Gunther 1990; Sapir 1915, 360–61.
14. McMillan 1999, 139; Drucker 1951; Coté 2002; Huelsbeck 1988; Reid 2015, 10–11.
15. Coté 2002; Drucker 1951; Renker and Gunther 1990. On the control of access to ritual knowledge, see Drucker 1951. For detailed accounts of whale hunts based on the memory of second-generation informants, see Curtis 1911, 19–34; Drucker 1951, 50–55.
16. The archaeologist David Huelsbeck used evidence from the archaeological site at Ozette village to argue that whale products accounted for as much as ninety percent of consumable protein in that village prior to contact (Huelsbeck 1994), but Huelsbeck's calculations could be misleading, since he states elsewhere that the Ozette data could not "be used to accurately rank the relative importance of whales" to subsistence (Huelsbeck and Wessen 1994, 14). Archaeologists have also pointed out in more general terms the intractable limitations of archaeological data for conducting comparative faunal analysis (McMillan 1999, 176). Isaac (1988) writes that without more direct ethnographic data, "we are left to the devices of archaeology in an area with exceptionally poor preservation of organic remains. . . . [E]ven the 'wet sites' [like Ozette B70], with their superior preservation of fragile organic remains, have yielded little information on food uses of plants" (7).
17. Huelsbeck 1988a, 1994; McMillan 1999, 137; see also Bowechop 2004, 407.
18. Colson 1953; Drucker 1951; Huelsbeck 1988b; Jewitt and Alsop 1815; McMillan 1999; Reid 2009; Swan 1870; Taylor 1974; Wessen 1982; Bowechop 2004.
19. Wessen 1982; Reid 2009, 52–55; McMillan 1999, 153.
20. Isaac 1988, 11; Renker and Gunther 1990; Reid 2015, 24–26.
21. Renker and Gunther 1990.
22. Drucker 1951, 11; Mozino and Engstrand 1991, 3, 67–71.
23. Arima and Hoover 2011, 79.
24. Reid 2015, 92.
25. Reid 2015, 92–105.
26. Reid 2015, 97; see also Harmon 1998.
27. Collins 1996, 183; Gibbs 1877.
28. Reid 2015, 98.
29. Reid 2015, 136; Collins 2012.
30. For a more comprehensive account of the negotiation and signing of the Treaty of Neah Bay, see Collins 2012.
31. On smallpox outbreaks, see Renker and Gunther 1990. On the Treaty of Neah Bay, see Gibbs 1877; Swan 1870; Coté 2002, 86. The egalitarian dictates included the abolition of slavery, which had the effect of undercutting the tripartite class system.
32. Colson 1953; Collins 1996.
33. Erikson 2002, 86–90.

34. Colson 1953; Reid 2015, 176. See also *A Gift From the Past*, a 2010 film directed and produced by the Makah Cultural and Research Center.
35. Colson 1953.
36. Collins 1996.
37. Webb 1988; Waterman 1920. Joshua Reid (2015) disputes the idea that Makah whaling virtually ceased in this period (173–74), and he attributes any drop-off in frequency of expeditions to the growing scarcity of gray whales, noting that Pacific whale populations were already in sharp decline as early as the 1850s due to the effects of European industrial whaling. Waterman's (1920) claim that Makahs "left off whaling" between 1860 and 1890 could be explained by his informant only remembering his own family's history. But James Swan, a sympathetic Indian agent who resided on the Makah Reservation for years and wrote extensively on Makah customs, observed plentiful gray whales off the shores of the reservation as late as 1878, and he lamented the Makah's indifference in no longer bothering to hunt them (Busch 1987, 131; Doig 1982, 166).
38. Reid 2015, 184–92.
39. Webb 1988; see also Reid 2015, 191; Russell 2001.
40. While such double binds are arguably endemic to the politics of cultural recognition where cultural exemptions are concerned, the problematic shape of the state's indigenous sealing exemption can also be traced to the particular discourses on cultural authenticity that pervaded the political and anthropological thinking of the time (and arguably remains deep-seated today; Raibmon 2005, 4–14).
41. Collins 1996; Reid 2015, 176.
42. In an interview conducted by the Makah Cultural and Research Center, a Makah elder told a story of her grandmother going out on a whaling expedition around 1900 (Makah Cultural and Research Center 2010). According to Coté (2002), the acceptance of a woman's involvement in a whale hunt signaled the breakdown of whaling as a traditional institution at that time. See also Russell 2001.
43. See, e.g., Bowechop 2004, 409.
44. Sapir 1915, 356–59.
45. Reid (2015) observes that prior to contact, most of the kin relations between distant villages were among ranked chiefs (45–46).
46. Harmon 1998, 76. The nascent regional chieftainships that Reid (2015) describes likely emerged in the wake of the disruption caused by smallpox epidemics and collapsed by the time Stevens arrived to negotiate the Treaty of Neah Bay. Isaac (1988) observes that the chieftainships that emerged in this time on Cape Flattery and on Vancouver Island probably resembled most closely the configuration of "tribes" in the contemporary juridico-political sense of the term, and it was the large-scale trade in furs and, later, in whale oil that made these configurations possible (11).
47. Harmon 1998, 137–39.
48. Harmon 1998, 180, 203–4). The Dawes Act of 1887 further eroded tribal sovereignty and coherence by moving land from ownership in common to individual allotments, thus making it what Harmon calls the "keystone of the [federal] assimilation program" (1998, 138).
49. Harmon 1998, 180–81.
50. See Harmon on the effect of the hops industry for Puget Sound tribes (1998, 199–200). Oral histories offered to Erikson (2002) mention the hops fields as a site where

bone gaming and other pantribal activities could flourish beyond the reach of Bureau of Indian Affairs agents assigned to particular reservations.

51. When the Northwest Federation of American Indians sued the federal government for failing to live up to the promises of the Stevens treaties in the early twentieth century, witnesses for the plaintiffs routinely testified that their knowledge was limited to their own family traditions (Harmon 1998, 185). Early ethnographers showed a lack of understanding on this point by purporting to write ethnographic representations of "tribes" based on data collected from just one or two family lineages. It is common today for Native peoples in the region to say that Franz Boas or Edward Curtis wrote an ethnography of the Smiths or Johnsons (whichever family their primary informants belonged to) rather than of the tribe in question.
52. Harmon 1998, 188.
53. Harmon 1998, 188–89.
54. Erikson 2002; see pp. 102–8 for relevant oral histories offered to Erikson.
55. If there is a disagreement within Colson's account regarding the vitality of tradition and the memory of whaling during this time, one possible explanation, again, is the cultural and spiritual diversity of family lineages within the Makah Tribe. By Colson's time, some families had converted to forms of Christianity that discouraged the retention of traditional Native practices (or at least discouraged people from talking about those practices with visiting outsiders). Which family Colson spoke with at any given time could have influenced her sense of the discursive importance of whaling pasts for "the tribe."
56. On tribal governments' lack of authority at the time, see Ziontz 2009, 65.
57. Ziontz 2009, 50, 53. The film is mentioned only in Al Ziontz's memoir *A Lawyer in Indian Country* (Ziontz 2009), and I should note that I can find no corroborating evidence for its existence. Ziontz was the tribal lawyer for the Makah for years, and in the memoir he admits his bias and sternly refers to opponents of Indian rights as "the bad guys." His factual claims should thus be viewed with a measure of suspicion; but the existence of such a film in the political climate of the time is plausible.
58. Harmon 1998, 186.
59. *Makah Indian Tribe et al. v. Schoettler, Director of the Department of Fisheries*, 192 F.2d 224 (9th Cir. 1951).
60. Coté 2002; Harmon 1998, 233; Sepez 2001, 68.
61. Ziontz 2009; Harmon 1998, 228; Coté 2002, 133–34.
62. Subsequent decisions have extended the Boldt decision to include other fish species. For the Makah, this has thus far meant an equal share in the harvest of halibut, herring, shellfish, and Pacific whiting.
63. *Winans* reversed a previous ruling by a district court in Washington that while the Makahs had "rights in common with all citizens of the United States," they had no "peculiar or superior rights or privileges denied to citizens of the country in general." *United States vs. The James G. Swan et al.*, 77 F. 473 (N.D. Wash. 1896).
64. Archaeologists had been working at Ozette for several years before the discovery of the entombed village. The earlier sites were what archaeologists refer to as "dry sites," which, due to the nature of decomposition, contain few if any organic remains. What the hikers discovered in 1970 was a "wet site," where encasement in moist clay had allowed for the preservation of artifacts that would have otherwise turned to dust.
65. The village of Ozette was the most distant, geographically and politically, of what would later be referred to in the Treaty of Neah Bay as the villages of the Makah Tribe

(McMillan 1999, 86; Arima and Hoover 2011, 23), but the local consensus today is that Ozettes and Makahs were the same (see Erikson 135–38). The Makah Cultural and Research Center and its collections are based almost entirely on artifacts recovered from Ozette. Archaeological research at the Ozette site has served as the basis for Makah legal claims related to whaling (Renker 1997, 2007).

66. For an engaging account of the excavation itself, see Kirk and Parker (2015). See also Bowechop (2004) for a reflection on the effects of the discovery for young Makah men and women.
67. Nor is accidental entanglement an uncommon occurrence today, although its precise frequency is difficult to estimate because many entanglements go unreported. Scordino, Carretta, and Cottrell (2014) reported thirty-five gray whale entanglements in fishing gear in the waters around northwest Washington between 2008 and 2012, with Erikson (1999, 556) reporting that five whales had been lethally entangled in Makah fishermen's nets each year between 1980 and 1995. These rates may be decreasing due to changes in fishing methods and increased efforts to mitigate marine mammal bycatch (Scordino and Maté 2011).
68. On the MCRC as a site for representing and performing Makah identity, see Erikson 2002, 18–30.
69. Kirk and Daugherty 1974.
70. Sepez 2001, 98–99. The term *party* is often used by Makahs to refer to potlatches, as is the case here. The "pile of seals" refers to seal meat that was prepared for the occasion.
71. E.g., Reid (2015, 1); Sepez Aradanas 1998; Van Ginkel 2004, 64; Erikson 1999; Sullivan 2000; Peterson and Makah Cultural and Research Center 2003, 163; *Anderson v. Evans*; Reid 2015. While there are many examples to choose from, Coté's (2010, 7) description is especially misleading: "When the Makah tribal leaders *heard* that the gray whale was no longer endangered, they announced that they were going to revive their whale hunts" (7, my emphasis).
72. See Beck 1996, as quoted in Kim 2015, 221; see also Martello 2004, 266; Russell 2001, 102; Deur 2000, 146. Peterson (1996) calls it "common knowledge" that the Makah, along with various other commercial fishing groups and treaty tribes, petitioned for the delisting, which makes the equivocation in so many other accounts appear motivated.
73. See Huelsbeck 1988.
74. As author Dick Russell observes, the Treaty of Neah Bay prohibits the Makahs from trading outside of the United States, and there is currently no domestic market for whale meat or blubber (2001, 103).
75. A Northwest Indian Fish Commission official reported in the April 1995 meeting of the Pacific Scientific Review Group that "the Makah intend to harvest gray whales (starting in 1996), harbor seals (5 already taken), California sea lions, minke whales, small cetaceans such as harbor porpoise and Dall's porpoise, and, potentially in the future, sea otters. The Makah are planning to operate a processing plant so as to sell to markets outside the United States. The Makah have started discussions with Japan and Norway about selling their whale products to both countries. The plant could be used to process the catches of other tribes as well."
76. Bill Frank Jr. to the Honorable Robert Mosbacher, 16 October 16, 1991.
77. William W. Fox Jr. to Bill Frank Jr., November 27, 1991.
78. John R. Twiss Jr. to William W. Fox Jr., May 15, 1992.
79. William Fox Jr. to John Twiss, December 30, 1992. In response to Twiss's concerns

regarding the uncertainty of gray whale population structure and breeding habits, Fox reminded Twiss that the Endangered Species Act (only) required that the agency take into consideration "the best available scientific and commercial data available."
80. Blow 1998; Russell 2001, 286. Sullivan (1998) reported the amount of the grant from the US Commerce Department at $310,000, as did Verhovek (1999).
81. Ross 1992.
82. Hubert Markishtum, Chairman of the Makah Tribal Council, to Will Martin, Deputy Assistant Secretary of Commerce for International Affairs, "Re: Makah Indian Tribe—Whaling Proposal," May 5, 1995.
83. Hannula 1995.
84. Hubert Markishtum, Chairman of the Makah Tribal Council, to Will Martin, Deputy Assistant Secretary of Commerce for International Affairs, "Re: Makah Indian Tribe—Whaling Proposal," May 5, 1995.
85. Renker 1997, emphasis added.
86. Renker 2007. If Renker's concept of cultural need had a precursor, it was the 1979 report of a panel on the cultural aspects of aboriginal whaling formed under the auspices of the International Whaling Commission, which found that hunting bowhead whales was essential and necessary to the preservation of Eskimo culture and mental health. See Scarff (1980, 274).
87. Described thus, Makah whaling shares some of the characteristics of a revitalization movement (see Harkin 2004, xxv-xvi). I have no particular objection to the use of that term in this case (outside of its tendency toward presuming linearity in social processes; Harkin 2004, xxx), I just do not find it useful as a model for understanding the complex and dynamic relationship between whaling and Makah identity.
88. Cote 2002; Ziontz 2009.
89. For an attempt at a balanced ethnographic analysis of the tensions around indigenous fishing rights among nonaboriginal Canadian fishermen, including anxieties specifically related to the Boldt decision and its outcome for US fishermen, see Menzies (1994).
90. "Ceremonial and subsistence" is a federal management classification created by *United States. v. Washington* in order to reserve tribal members the right of taking fish for "personal subsistence consumption" and "traditional tribal ceremonies."
91. See Sullivan 2000, 56-59.
92. Coté 2010, 135.
93. See Ziontz 2009, 256-57.
94. Ziontz 2009, 257-60.
95. See Sullivan 2000.
96. Coté 2002, 145.
97. See Harmon 1998, 232; Coté 2010, 115-21.
98. The kind of protest that Jerry Lucas advocates bears a closer resemblance to the politics of refusal described by Simpson among the Kahnawà:ke (2014).
99. Mapes 2002.
100. See also Reid (2015) for a description of the celebrations following the 1999 hunt that stressed the importance of treaty rights and tribal solidarity, including an appearance by Billy Frank Jr.
101. See, e.g., Reid 2015, 2-3. I should clarify that while the event resembled a potlatch and has been informally referred to as such, potlatches are hosted by individual families, and this event was hosted by the tribe (see Bowechop 2004, 417). This is another way in which the "tribal" character of the 1999 hunt represented an exception, and

perhaps an expansion, of the horizon of possibilities for actions undertaken by the Makah Tribe as such.
102. Ziontz 2009, 264.
103. See Gibbs 1877.
104. See Erikson 2002, 187.
105. One of those "other things," according to the late Makah elder Charles "Pug" Claplanhoo in an interview with author Dick Russell, was the Claplanhoo family's last whale, harpooned in 1909 for the Pacific Exposition in Seattle. Claplanhoo also remembered that whale as the last whale the tribe brought in (Russell 2001, 97–100).
106. Markell (2003, 23). See also Butler 1999, 7.
107. Richard Dougherty (2001) quotes Theron Parker: "I'm gonna be a whaler. I didn't really have a choice in the matter, I guess. It's in the blood."
108. Sometimes genetic inheritance has been the literal terrain on which rights have been claimed on behalf of Makah whaling families as well. Ann Renker suggests that children from traditional whaling families may have a greater nutritional need to consume marine mammal fatty acids, citing as corroboratory evidence the greater prevalence of health problems and learning disabilities among children of whaling families. Renker has made this suggestion with increasing confidence in each version of the tribe's needs statement to the International Whaling Commission since 1997 (Renker 1997, 32; 2007, 60; 2012, 79; see also Coté 2010, 196–97).
109. For instance, as Patricia Erikson recounts, without the commitment of presiding archaeologist Richard Dougherty it is uncertain whether the site and its artifacts would have been transferred under tribal control in the first place (2002, 140).

CHAPTER TWO
1. See Bock (1995) for a journalist's account based on interviews with the fisherman, Dan Greene, and several others involved in the butchering of what Bock called "the accidental whale."
2. Halibut and salmon, two of the most commonly consumed fish in Neah Bay, cost between $15 and $25 per pound in Seattle markets, yet the monetary value of these foods is not factored into official estimates of income, nor do subsistence-gathering activities qualify as employment in the calculation of the unemployment rate.
3. Minus tides are low tides where the water recedes past the annual average. Neah Bay has one of the largest intertidal zones in North America, which attracts marine ecologists and beachcombers alike.
4. See, e.g., Boas 1897; Drucker 1951; Renker and Gunther 1990; Swan 1870; Waterman 1920.
5. I'm indebted to Harriet Ritvo for pointing out the silliness of the word, which I had stopped noticing.
6. The *Peninsula Daily News* is a regional newspaper that serves the northern Olympic Peninsula. The paper's circulation is approximately 14,000 on weekdays and 15,000 on weekends.
7. Quoted in Russell 2001, 114.
8. Peterson (1996) reported this at the time.
9. Kim 2015, 240. I heard this story repeated by a number of people, although it was sometimes reported that her cat was killed.
10. This phrase, "Take only what you need," is used commonly enough that Renker included it in the Makah needs statement (1997) as a pithy statement of what she called the community's subsistence strategy.

11. I am uncertain when and where the Thunderbird illustration itself first appeared, but it was already a meaningful image as early as the 1850s, when James Swan observed a version painted on a board in a chief's lodge in Neah Bay (Swan credits a Clayoquot artist with that particular painting; 1870, 7–9).
12. Based on an interview with Luke Markishtum in the 1930s, anthropologist Erna Gunther reported that killer whales were "killed for the pleasure of the chase, for the meat was not used" (1936, 117).
13. This interview made me think of Donna Young's observation in her essay "Writing against the Native Point of View" of how our voices as anthropologists appear in the text "newly altered so as to be measured and thoughtful," while the voices of our interlocutors "remain raw," a tricky situation—Young thinks of it as a kind of betrayal—that has "absolutely nothing to do with obtaining consent" (213). Feeling uncertain about committing such a contrast to print, I called Jo Jo to make sure he was comfortable with my reporting of the interview. I read him some of the quotes and told him my concerns. "If that's what I said, that's what I said," was his reply. He seemed to have no problem whatsoever with my memory or my portrayal of events. That has not shaken me from the feeling Young describes—not a betrayal, but a concern that neither participant in this interview ever fully understood its meaning for the other.
14. Robert Sullivan called Micah McCarty a "a kind of Save the Whales whaler, who saw environmentalism and whaling as compatible if conducted on ancient Native American terms" (2000, 38). Dave Sones expressed a similar gratitude toward the save-the-whales movement, which he saw as beneficial to Makah whalers who now wanted to "reintroduce whales in our culture." While Sullivan's description of Micah had an element of truth to it, Micah had little patience for what he called the pious dogma of antiwhaling "eco-Evangelists." He once told me that attending an Earth Island Institute conference made him feel like "an atheist at a Jehovah's Witness convention."
15. For a well-known and thorough review of this problem in ethnographic and historical accounts, see Krech 2000.
16. On charges of inauthenticity specifically related to Makah whaling, see Deur 2000.
17. Some versions of this argument are nuanced enough to transcend the question of whether hunter-gatherer beliefs about animals are literal or figurative, presenting a picture of the complex and blurred boundaries between subject and object, spirit and body. Rane Willerslev (2007), for instance, provides an account of Yukaghir hunters of northeastern Siberia, for whom animal personhood "is not explained as a property of the animal as such, but of its position with the relational field of mutual mimicry" (73–118). "The animal's personhood," Willerslev argues, "results from the coming together of the hunter's own awareness as a person and his experience of the prey animal acting as a reflection of his own body" (118).
18. On the inherent shortcomings of Nadasdy's somewhat fanciful suggestion, Harkin writes: "Such a philosophy of the world cannot, and never could, provide a road map of resource management, as some authors have supposed. . . . Rather than a road map, they provide a poetics; rather than technology, the consolations of philosophy" (2007, 229).
19. On the question of reciprocity in human-animal relations, see Harkin (2007), who writes of the tension produced by a cosmology where humans never quite repay their debts, and "what appears at first glance to be a system based upon reciprocity relations turns out to be more fundamentally predatory" (222).

20. "Rez dog" is the common term used by members of the community to refer to the packs of dogs that roam the reservation.
21. See Rappaport (1968) on the Tsembaga (New Guinea) pig husbandry for an anthropological example of how animals can be beloved *and* eaten.

CHAPTER THREE

1. See Howe (2016) for an analysis of how American Constitutional law presupposes a Protestant model of religiosity that is a poor fit with the material, place-based cosmologies of many Native American groups.
2. I would emphasize "contemporary" and "public" here, since Waterman (1920) noted many years ago that while Makahs "see with the very greatest care to all practical matters connected with whaling expeditions . . . they seem to look upon the religious performance as much more essential to success" (38); and a number of Makah whalers told me they were taught as children to prepare spiritually decades before the current revival.
3. Commercial whaling in the United States is prohibited by the Whaling Convention Act. See also Russell (2001), who notes that the Treaty of Neah Bay prohibits the Makah from international trade and that there is no legal market for whale meat within the United States (103).
4. Judge George Boldt created the "ceremonial and subsistence" category in his ruling in *United States v. Washington* in order to reserve tribes the right of taking fish for "personal subsistence consumption" and "traditional tribal ceremonies."
5. On the slippery slope, see, e.g., Will Anderson's 2006 essay, in which he writes: "IWC approval of Makah whaling would have a profound effect on other whales (there are also thirteen tribal Indian bands in Canada and an Alaska Eskimo Whaling Commission that have stated their intent to kill gray whales). Whale protectionists familiar with the IWC know that the biggest beneficiaries of a Makah IWC victory would be the Japanese, Norwegian, and other commercial whalers" (387). See also Russell (2001, 100–102, 104) on Japanese commercial whalers' interest in leveraging cultural whaling to bolster support for their own "small-type coastal whaling" operations.
6. Native American religious practices are also protected by the American Indian Religious Freedom Act of 1978.
7. Since traditional and spiritual knowledge in the Makah community tends to be transmitted within extended family lineages and remain peculiar to that family, any detailed views provided by informants in this chapter should be taken as representative of that person's individual attempt to make sense of things or, in the broadest sense, as representative of his or her family.
8. Coté 2010, 50–52.
9. For her discussion of spiritual power, see Colson 1953, 249–60.
10. In a moment of analysis that is perhaps problematically Protestant in its assumptions about how conscience works, Colson said of the tenacity of traditional belief, "You may ban the manifestations, but you cannot directly affect the theories or beliefs or attitudes which underlie the behavior and which give the behavior its meaning to those who practice it" (1953, 293).
11. See Erikson (2002, 102–8) for some Makah elders' reflections on the different character of Neah Bay churches and their influence on the preservation of non-Christian customs.
12. See chapter 4.
13. The picture of the whale as a transcendent spirit and each individual creature as

a single instantiation of it fits with Timothy Ingold's original observation—which he has subsequently retracted—that North American hunting peoples do not view animals themselves as persons but only their higher-ranked spiritual types (1987, 247).

14. E.g., "Were Native peoples the ones who decimated the whale populations in the first place? No. We have always lived in harmony with the whale and with nature, and our sustainable use of our resources and our utilizations of the environment has always been conducted with respect" (Coté 2010, 192). Krech (1999) notes that one of the first architects of the "noble ecological Indian" was the Dakota Sioux intellectual Ohiyesa, aka Charles Eastman (19–20).
15. Several examples come to mind that demonstrate this opinion, but there is one I remember most vividly. After gathering stinging nettles along with a group of Makahs at an event led by a community health nurse, I made the mistake of confusing a fishermen with another individual who was working on a project making traditional fishing line for the museum (neither of whom were present). The Makah woman I was speaking with quickly corrected me, saying that the fisherman probably wouldn't be interested. "He is a fisherman," she said. "They always take the easy way out."
16. Collins 1996.
17. Bowechop 2004.
18. Technically, the whaling commission guidelines stated that "the responsibility of the Whaling Captain is to see that each crew member . . . is adequately prepared spiritually and physically to go on a hunt" (Sullivan 2001, 203). The Makah Whaling Commission was modeled after Alaskan indigenous whaling commissions, where family-based hunts led by patriarchs and an otherwise egalitarian social structure make such an edict more realistic and effective. In practice, Wayne Johnson lacked the familial authority and expertise and arguably also the public stature to execute such a command. The full list of the whaling crew certification qualifications are as follows (from Sullivan 2001, 202):

> Each member must be able to swim in the open ocean.
> Each member must pass a Drug and Alcohol test.
> Whaling crew must have expertise in shooting the rifle on land, in canoe, and practicing on a moving target.
> Whaling crew must have experience in use of harpoon on land in canoe, and practicing on a moving target.
> Must be able to show that they have made attempts at approaching a Whale in a canoe.
> Must be able to tread water.
> Must be cross-trained in canoe.
> Chase/Support boat crew must be able to drive boat in open ocean, surface, etc.
> Chase/Support boat crew will be tested for Drugs and Alcohol.

19. Some members of the 1999 crew either failed drug tests or refused to take them in the time leading up to the 1999 hunt, which is what McCarty appears to be referring to (Sullivan 2000, 212–23).
20. Bock 1995.

CHAPTER FOUR

1. When Makah leaders are being especially careful, they employ the somewhat cumbersome term "reserve-right treaty right" to emphasize that Makah rights preexisted

the treaties according to US common law, a Supreme Court precedent known as the Reserve Rights Doctrine.

2. Makah whaling is permitted by NMFS as a "ceremonial and subsistence" (C&S) hunt, a category defined by the Boldt decision (*United States v. Washington*) in order to exempt certain catches from the calculation of the total harvest for the purposes of dividing annual catches between tribal and nontribal fisheries. C&S catches may not be sold commercially; as such, they constitute a juridically produced version of what Kopytoff (1986) calls a distinct sphere of exchange. By excluding capture for sale, the C&S category defines cultural or traditional activities in opposition to trade and commerce, which reinscribes the problematic understanding of culture as outside of economics. Because C&S exemptions are not tethered to specific methods or modes of resource gathering, however, the category promises, at least in theory, a space of pure exception that is otherwise free of debates about authenticity (cf. Muehlmann 2009). In practice, however, C&S catches are subject to federal approval according to the principles of modern conservation and fisheries management, which is why I do not mention the classification much in the analysis that follows.
3. Jasanoff 1990; Jasanoff 2004, 280.
4. Thornton and Hebert 2015, 370.
5. *Traditional ecological knowledge* is a somewhat problematic and contested term. Some question whether it is really "knowledge" at all (Nadasdy 2004), while others criticize the use of "tradition" for connoting backwardness (Berkes, Colding, and Folke 2000, 1251).
6. On the modern state's efforts to simplify nature and human populations in order to better control them, see Scott (1998).
7. Nesper and Schlender 2007, 293. See also Stacey et al. 2012; Menzies and Butler 2006. For a recent story involving Makah halibut hook technology, see Ryan (2018). In one famous case described by Wohlforth (2005), Alaskan Inupiat whalers successfully mobilized traditional ecological knowledge to argue for revised population estimates of the endangered bowhead whale. In 1977, the International Whaling Commission estimated the population of the Bering stock of bowhead whales to number between six hundred and two thousand—too few to support a continued harvest. Inupiat whalers objected, arguing that the new estimates failed to account for whales that moved under the sea ice rather than along it (18–21). Although it took nearly a decade and required innovations in hydrophone technology to substantiate the claims, population estimates were eventually revised upward to 7,200 individuals in 1985. A limited hunt has been permitted since that time with no measurable effect on the total population.
8. Hunn et al. 2003; Ranco 2007.
9. Simpson 2014, 20; Coulthard 2014. See also Deloria 1995.
10. The first such organism Paine described was a sea star living in the intertidal zone of Makah Bay.
11. Although he does not come out and say it here, Micah implies that Makah whalers are needed to keep gray whale stocks in check. Makah officials pointed to a significant die-off of gray whales in 1999 and 2000, enough for NMFS to deem it an "unusual mortality event"—as evidence that the stock was overpopulated and in need of culling. Activists used the same event to caution against a hunt given the fragility of the stock amid changing ocean conditions.
12. See Colson (1953) for an analysis of the effects of assimilation programs on the Makah community up to that point in time.

13. Ziontz 2009, 126.
14. Sones makes this statement in *Makah Nation: A Whaling People*, a film produced by the Makah whaling commission. He shared a similar sentiment with me in an interview.
15. Robert Brightman (1993) describes how spiritual beliefs often required hunters to kill every animal they saw, lest they offend the animal's "offering" or lest the escapees warn other animals away from the hunting ground. Krech (2000) gives the example of Plains Indians who believed buffalo herds came from somewhere under the earth in inexhaustible numbers, making limitations on the hunt unnecessary.
16. For an exploration of the myth of the ecological Indian and its possible historical bases, including several critiques of Shephard Krech's original argument in *The Ecological Indian*, see Harkin and Lewis (2007).
17. Kaste 2014.
18. The Bald and Golden Eagle Protection Act expressly exempts the killing of bald eagles for the "religious purposes of American Indian tribes" through a permitting process that considers "the direct or indirect effect which issuing such permit would be likely to have upon the wild populations of bald or golden eagles" (50 C.F.R. §22.22[c1]), thus injecting a conservationist procedure into the heart of one of the country's few preservationist laws.
19. One can certainly dismiss this utilitarian ethos as a corruption of modern capitalism, as Coulthard (2014) suggests; but it is impossible to do so without suggesting that this former fisheries manager and the rest of Makah leadership is somehow *fallen* (see, e.g., Hogan 1995). Is it possible that Makah attitudes have changed in the last century, perhaps along the lines of the shift toward extractive capitalism that Coulthard (2014, 78) describes in the Dene Nation's attitudes toward land use? Of course. But as long as one is not *criticizing* Makahs for their utilitarian attitudes, this has no bearing on our understanding of the Makah whaling conflict in the present, in which only those Makahs currently living occupy positions of power and influence in the tribal government.
20. See the National Marine Fisheries Service's statement "Ecosystem-Based Fisheries Management" on its Office of Science and Technology website: https://www.st.nmfs.noaa.gov/ecosystems/ebfm/.
21. I am referring here to the moral economy of resource use inherent in the structure of the law and its implementation (as well as conservation more generally; see Hays 1959), but ecologists have also noticed that the employees of natural resource bureaucracies tend to have more utilitarian attitudes toward animals than the general public (see Peyton and Langenau 1985). By "utilitarian," I mean a set of attitudes described by Kellert (1980) in which "the relevance of animals is largely defined by their usefulness to people."
22. See the Scientific Committee Report for 1998, IWC SC/54/BRG7. The figure Keith cites falls within the range of sustainable takes determined by the IWC Scientific Committee in 2002 and is actually less than the potential biological removal of 575 that NMFS had calculated for the Eastern North Pacific gray whale in its most recent stock assessment report at the time.
23. McCarty 2003. See also Coté 2010, 164. Discrediting the claims of antiwhaling preservationists in this fashion is not a new tactic; John Laurence McHugh, a fisheries biologist and the head of the 1964 US delegation to the IWC, denounced calls for restrictions on the Antarctic whale fishery by opposing "aesthetic" objections to the

"more realistic position" that economic objectives mattered as well (Burnett 2012, 510–11); echoes of the derision of "nature lovers" in the early days of conservation by Gifford Pinchot and his associates (Hays 1959, 196–97).
24. Bowechop 2004, 418; see also Russell 2001. Based on an interview with Theron Parker, Russell (2001) reported "someone from the Coast Guard" called Theron at two o'clock on the morning of the successful hunt to tell him that Sea Shepherd had sailed to Friday Harbor to load more boats and volunteers and would not be back in time to contest an early-morning hunt.
25. Lavigne et al. 1999, 11; Kellert 1979, 108, 112.

CHAPTER FIVE

1. See also Freeman et al. 1998; Kalland 2009. Niels Einarsson calls antiwhaling environmentalists "as naïve as [his] 7-year-old" for worrying about environmental problems without realizing that people's livelihoods may be at risk (1993, 82).
2. See, e.g., Marker 2006; Cote 2010, 150; Reid 2015, 3. Claire Jean Kim has also noted the tendency of the Makah Tribal Council and its supporters to essentialize all opposition to Makah whaling as racist or imperialist (2015, 243). Indeed, the tribe's tendency to associate animal rights opposition with older, plainly anti-Indian sentiment is arguably one of the factors driving the response to outside opposition that I described in chapter 2.
3. The *Seattle Times* referred to the Sea Shepherd boat as "only the most physical manifestation of the opposition the Makah face" (Westneat 1996).
4. On the expanding criminalization of animal activism in the United States, see Potter (2013) and Greenwald (2017).
5. Mapes 2002.
6. As Burnett (2012) observes, "popular interest [in whales] cannot be said to have been very great in these years. . . . As for the notion of these creatures possessing 'intelligence' or 'beauty,' one must comb closely through marginal materials to catch even a hint of such preoccupations in the later 1950s" (522). There was even talk of converting Pacific atolls into "whale farms" (523), very far indeed from what Burnett calls the extraordinary whale-hugging enthusiasms of the 1970s (524).
7. Epstein 2008, 46–50. See also Robertson 1954, 41.
8. Burnett 2012, 325, 351–52.
9. See Burnett (2012) on what he calls the shift in "world opinion" on whales and whaling (328, 458).
10. See Moore 2008, 58–59.
11. Lilly's research heavily influenced Paul Spong, a key intellectual figure in the Don't Make a Wave Committee, the precursor to Greenpeace (Burnett 2012, 639–42).
12. Makah elder Jean Vitalis told me a "horror story" involving gray whale breath. She was out on a fishing boat with her brother when he tricked her into looking over the railing. A gray whale sprayed her with a "godawful geyser" right in the face. "Sand and shells and slime—it was the most disgusting thing," she said, "and it *burned.*"
13. In a noteworthy exception, Chuck recalls with regret killing hundreds of unwanted dogfish by slitting their bellies and throwing them back over the side of the boat. Dogfish are small sharks that are loathed by commercial and sport fishermen for their aggressive behavior and penchant for mangling fishing tackle. "We just didn't even think about it," he said.
14. For a similar sentiment, see Peterson and Hogan (2003), who write: "The great

whales . . . are beyond our rights, our treaties, and our histories. They exist beyond our own needs, in a world of land and sea that we share, but that we have never truly fathomed and never owned" (xvii).

15. Arne Kalland (2009) points out that even as whales became icons of the environmental movement, few Americans actually encountered them in real life. As a result, prevailing public conception of whales corresponds to no single species of whale or dolphin. Kalland suggests that antiwhaling activists deliberately promoted this "super-whale." "We are told that the whale is the largest animal on earth," Kalland writes (this applies to the blue whale), "has the largest brain on earth (the sperm whale), has large brain-to-body weight ratio (i.e. is intelligent, the bottlenose dolphin), has a pleasant and varied song (the humpback whale), is friendly (the gray whale), is endangered (the bowhead and blue whales)" (29). Kalland's close professional association with Japanese whaling companies paint his motives in a questionable light, but there is truth to his observation about the tenacity of the image of the superwhale. "We are talking about the largest animal that ever lived, the largest heart that ever beat!" Margaret once said to me of gray whales, neither of which is true of the species.

16. Chuck wrote a letter to the editor of the local paper pledging to the Elwha Tribe that "your friends and neighbors will lie down beside you in front of those bulldozers." Four days later, according to Chuck, the project was halted.

17. Placing the heads of vanquished enemies on sticks in front of one's house is a widely documented act of retribution in Northwest Coast warfare (see, e.g., Drucker 1951).

18. Sullivan 2000, 96.

19. Zodiacs are small, rigid inflatable boats favored by Sea Shepherd and other antiwhaling organizations.

CHAPTER SIX

1. The authors of the study (Lang et al. 2011) included representatives of the National Marine Fisheries Service, Cascadia Research Collective, the Makah Fisheries Marine Mammal Program, the Alaskan Department of Wildlife Management, and Oregon State University's Marine Mammal Institute, as well as representatives of Russian indigenous whaling interests.

2. See Scordino, Bickham, Brandon, and Akmajian 2014.

3. The full summer range of the resident/PCFG whales is estimated to extend from northern California in the south to southeastern Alaska in the north. The estimate is based on photographic identification records.

4. Specifically, the study found statistically significant differences in the relative occurrence of mitochondrial DNA (mtDNA) haplotypes between individuals sampled from the resident/PCFG subpopulation and individuals sampled from the rest of the Eastern North Pacific stock. No nuclear DNA differentiation was found. These findings support the hypothesis that "recruitment" of the PCFG stock is exclusively driven by calves "learning" the feeding locations from their mothers, a process that scientists refer to as "matrilineally directed site fidelity" (Lang et al. 2011, 3). The authors wrote: "If recruitment into the PCFG is driven exclusively by the return of individuals which followed their mothers to the area as calves, then over time those mtDNA haplotypes originally found only in males or non-reproducing females would be removed via genetic drift, while haplotypes found in females and their returning offspring would build to higher frequencies" (7). The finding significantly impacts

management decisions regarding the stock because it suggests that a resident/PCFG whale that is "removed" (killed) would not be replaced by "immigration" from other feeding groups (3).

5. "If the intent is chasing down mother-calf pairs to try to trace their mitochondrial DNA, yes, you are going to find something," Micah told the *Peninsula Daily News* (Gottlieb 2012) after clarifying that "there is a degree of validity. . . . I'm not saying [the findings] are incorrect." He has since pointed out, however, that "resident whale" is a misnomer since there is some turnover in the feeding group from year to year (Coté 2010, 174).

6. In comments on a previous version of this manuscript, a representative of the Makah Fisheries Department pointed out that resident/PCFG whales are not considered a separate stock by the National Marine Fisheries Service. Indeed, a 2013 report commissioned by NMFS indicated that the question of whether the PCFG represented a separate stock is "yet to be resolved" (Weller et al. 2013), although NMFS has stated elsewhere that "PCFG whales may warrant consideration as a separate management unit" (Golde 2012), and NMFS's 2014 stock assessment calculated a potential biological removal for the PCFG group as if it were a distinct population segment. Technically, the current indefinite suspension of Makah whaling is due to delays in the MMPA waiver process and in the satisfaction of other requirements set by the Ninth Circuit Court in *Metcalf v. Daley* and *Anderson v. Evans*, which delays are the result (at least in large part) of the new scientific findings on resident/PCFG whales and other gray whale populations. My point here, however, is that the situation follows the general pattern of weak-stock fisheries management and that the domestic gray whale catch quota for Makah whalers is effectively, if not officially, zero.

7. See Burnett 2012 for a complete history of the IWC Scientific Committee and its agendas.

8. Burnett 2012.

9. See Burnett 2012, 81.

10. Scott McVay called the cetology of the early to mid-twentieth century "parasitic"; see Schevill, Ray, and Norris 1974, 381; also quoted in Burnett 2012, 188.

11. Gilbert and Dodds 1987; see, for instance, Leopold 1933. The use of agronomic metaphors in fisheries management goes well beyond the more common (and obvious) terms like *harvest* and *yield*. The Makah National Fish Hatchery's website, for instance, describes the process of harvesting eggs from Chinook and Coho salmon as follows: "Adult fish are checked to determine overall numbers and 'ripeness' of eggs. When enough of the fish are ripe, they are crowded into the sorting and spawning area. Ripe fish are quickly killed to facilitate handling while 'green' fish are sent back to the holding pond through a waterslide system."

12. On the Prussian and Saxon origin of modern scientific forestry, see Scott (1998, 14). See also Kosek (2006, 91) on how early American wildlife management represented a "direct borrowing" from forestry management.

13. Because the agency usually does not respond to claims it considers irrelevant (see Stone 2015), it is rare to find a statement from a NMFS official that expressly dismisses ethical claims about whales. In one such case, Lynda Mapes (2002) recounted her conversation with NMFS fisheries biologist Pat Gearin in the *Seattle Times* in which Gearin explained that the delay of whaling was "a political decision that science couldn't support. It was some politician's interpretation, to avoid the so-called 'friendly whales'. But there really was very little science behind it." Mapes re-

ported: "No conservation issues will be raised should the tribe take another whale this spring, Gearin said."

14. For more on the ways that authorized technocratic discourses work to marginalize alternative perspectives, see Jasanoff 2004; on the same in the context of debates over nuclear energy, see Wynne 2010.

15. In the animal rights movement, aesthetics has long been one of the few recourses available to activists who wish to seek legal standing to bring charges on behalf of animals. In what legal scholar Christopher Stone (2010) calls the "tortured reasoning" of the courts on this matter, judges have allowed activists to bring suit on behalf of animals based not on the animal's suffering but on the aesthetic injury purportedly done to the activist who witnesses the suffering (65–66). In a lawsuit based on the Animal Welfare Act, for instance, the Animal Legal Defense Fund was only able to gain standing to challenge a USDA regulation for the treatment of captive primates by claiming that the plaintiff had suffered "aesthetic and recreational injuries" upon seeing the physical and emotional condition of captive primates (2010, 37–38).

16. Stock assessments from 2002 to 2011 included just a single paragraph on the putative "pacific coast feeding aggregation" and contained the following statement: "While some animals in this group demonstrate some site-fidelity, available information from sighting records . . . and genetics . . . indicates that this group is a component of the eastern North Pacific population and is not an isolated population unit" (e.g., Allen and Angliss 2011). The 2013 stock assessment, the first published after state-sponsored genetic studies replicated earlier findings the activists had promoted, expanded to eight full paragraphs and noted that "the status of the PCFG as a population stock remains unresolved" but that, as previously noted, it "may warrant consideration as a distinct stock in the future" (NMFS 2013).

17. Another wrinkle was added to the story in 2010, when Russian oil and gas interests funded a study by Oregon State's Marine Mammal Institute to tag and track a critically endangered Western North Pacific (WNP) gray whale in the hope of locating the breeding ground for the stock. The WNP population was thought to be completely demographically isolated from the Eastern gray whales at the time, but their breeding grounds were unknown to scientists. To the great surprise of everyone involved in the study, a Western gray whale nicknamed "Flex" traveled from his feeding ground near Sakhalin Island directly across the Pacific to the coast of Alaska, mixing with individuals from the Eastern North Pacific stock. The scientists then tagged a second Western gray whale, nicknamed "Varvara," who proceeded to undertake the longest known migration by any mammal in the history of science, traveling from Sakhalin Island to the Eastern North Pacific gray whale's breeding grounds in Baja California, and back. Forced to account for the possible presence of the Western stock in the Makah hunting area, NMFS calculated a potential biological removal for the Western gray whales (which includes accidental kills like fishing gear entanglement and ship strikes) at .06, or one whale every seventeen years. To the relief of tribal whalers, NMFS scientists quantified the risk that the Makah will "take" a critically endangered WNP whale at one whale every 83–167 years. Skeptical of the models used to quantify the risk, activists see in the Western stock a new potential avenue of blocking the hunt.

18. Describing the challenge, the task force observed that "most biological 'populations' and 'stocks' do not exist as truly distinct groups, nor are individuals within the same population typically part of a truly panmictic group. Rather, population

differentiation occurs along a continuum, and placing discrete boundaries along this continuum for management purposes is a challenge" (Weller et al. 2013, 8). Since the early 2000s, population genetics has provided the primary basis for stock differentiation—hence the importance of the discovery of differences in mitochondrial DNA haplotype frequency without nuclear DNA differentiation among the PCFG, which suggested that internal recruitment via matrilineally directed site fidelity exceeded external recruitment for the group. However, Frasier et al. (2011, 45) have questioned the use of genetic data as the sole criterion for stock differentiation, noting that "populations can be demographically independent even when long-term migration rates are high enough to result in relative genetic homogeneity." The 2013 task force suggested that *demographic independence* better captured the agency's operational criterion for distinguishing stocks than demographic isolation, since the latter implied complete reproductive isolation while the agency had, for some time, formally defined as stocks those populations where the "population dynamics of the affected group is more a consequence of births and deaths within the group . . . rather than immigration or emigration" (Weller 2013, 4). Although the report is too lengthy and substantive to summarize here, the task force raised a number of questions, including whether it is permissible for some females and males to interbreed with individuals from other stocks, and whether, say, a single orca pod could count as a stock, since it arguably constituted a "functioning element of the ecosystem" (a key component of the MMPA definition of a stock) despite extensively interbreeding with other pods.

19. There is a long history of stakeholders leveraging scientific uncertainty in debates about the conservation of large cetaceans. As Graham Burnett (2012) points out, prowhaling nations arrived at every IWC meeting from the 1950s to the early 1960s with their own set of scientific findings showing that Southern Ocean stocks may not, in fact, be collapsing. In the 1990s the tables turned, as Burnett quotes an exasperated Norwegian economist at the IWC who had exactly these sorts of tactics in mind when he wrote that "the new wave of anti-whaling scientists . . . were deploying scientific uncertainty speciously to delay action they objected to on fundamentally nonscientific grounds" (352; this was in response to activist efforts to quash the IWC's Revised Management Plan, which would have ended the global whaling moratorium and allowed limited commercial whaling).

20. In the terms introduced by Scarff (1980), which classify ethical arguments regarding cetaceans into three main categories, activists have moved from arguing about "the ethical propriety of killing cetaceans" to "the morality of risking or actually causing the extinction of species or smaller taxonomic units" (243).

21. It is also important to remember that the activists' victories in the Makah whaling conflict only apply to a single Native American tribe. Without a dedicated activist opposition, Alaskan Inuit whaling has continued without significant controversy and with little fanfare during the same time period.

CHAPTER SEVEN

1. Sperm whales have recently figured out how to gingerly pluck halibut off longline hooks as Makah fishermen reel them in, and the fishermen regard this as an expensive problem. I overheard a Makah fisherman explaining the behavior to tourists at a local restaurant: "What is it they say, if you see a sperm whale even once in your lifetime, you're very lucky?" He turned to his fellow fisherman to say, "How lucky are we, then? We see like five a year!"

2. See Peterson and Hogan 2003, 266.
3. On the archaeological data and the consumption of different whale species by Makah ancestors, see Huelsbeck (1988). According to Huelsbeck, among the whale bones collected at the Ozette archaeological site that were identifiable by species, about fifty percent came from humpback whales and forty-six percent from gray whales.
4. See Russell 2001.
5. Horwitz 2014, 90.
6. Along these same lines, the Makah Whaling Commission's list of qualifications for whaling crew certification, drafted by veterans of the tribe's fisheries department and reproduced in Sullivan (2000), mandates that potential crew members "must be able to show that they have made attempts at approaching a Whale in a canoe" (202) but offers no guidance in cases where a whale approaches a crew member.
7. An environmental assessment is not a public service announcement—it is not intended to educate or inform the public on anything other than the agency's assessment of the potential environmental consequences of a proposal—which is why I think the agency's didactic final warning should be read as exasperation.
8. For an overview of the state of this art, see Whitehead and Rendell 2014.
9. According to a 2011 study, killer whales kill up to one-third of all newborn gray whale calves each year as they migrate northward along the coast of California (Barrett-Lennard et al. 2005).
10. For an exception, see Anderson 2004. Pearson and Weismantel (2010) stress the importance of examining how agency has been historically defined rather than simply treating it as a "transcendent feature of being" (27).
11. As Anders Blok (2007) notes, "What ANT [actor-network theory] does is to drastically reduce the usual requirements for social action capacity . . . leaving a minimal requirement for actants to induce some contingency in social interaction" (73).

CONCLUSION

1. A basic description of the whale's behavior was immediately available to the public in the account that Wayne Johnson offered to a reporter (Mapes 2007).
2. Anderson distributed a pamphlet under the label of the Progressive Animal Welfare Society prior to the 1999 hunt in which he wrote: "Gray whales . . . are well-known for approaching people in a curious and friendly manner. When the Makahs move in for the kill, the whales will be unafraid" (quoted in Sullivan 2000, 16–17).
3. Ollikainen 2009.
4. Marshall Sahlins has made the important point that some postmodern critiques overlook the fact that "not all differences in practice are necessarily differences in form—let alone negations thereof" (2013, 10).
5. There *is* a way in which the next legal decision on Makah whaling could turn on something other than stock-based analysis, however. In a passage of the opinion in *Anderson v. Evans* that has received very little attention, the court expressly rejected the notion that summer resident (PCFG) whales possess value only as genetically distinct stocks, writing that the local "environment" could still be affected even if the PCFG stock was not significantly impacted in its entirety. The administrative law judge (ALJ) who reviews the final EIS might conclude, as activists have argued, that changes in the behavior of summer resident whales resulting from annual hunts would count as a significant impact on the local environment even if the PCFG stock remained healthy and numerous in the waters off Oregon and Alaska. Conversely,

the ALJ could rule, in keeping with the National Environmental Policy Act's stated objective to "preserve important historic, cultural, and natural aspects of our national heritage," that the local environment has historically included Makah hunting activities and ought to include it in the future regardless of its potential impact on the population dynamics of any one gray whale stock.

6. Japan's announcement of its withdrawal from the IWC (as this book was going to press) represents the first major reversal in the perceived progress of the antiwhaling agenda since the moratorium on commercial whaling went into effect in 1986, and many activists will likely see this development as a consequence, in part, of the United States' inconsistent position on whaling viz. Alaskan and Makah whalers (see p. xxx, n. 5).
7. Cote 2010, 187–88.
8. See, e.g., Descola 2013; Kohn 2013; Ingold 2000; Nadasdy 2007.
9. The American institution of slavery represents an exception but not a contradiction, according to Kopytoff, since this separation of people and things is precisely *why* slavery presented "an intellectual and moral problem in the West . . . but almost nowhere else" (1986, 84). This is the moral denotation that is lost if we call whales "nonhuman persons" for those who hunt them.
10. See p. xxx, n. 6.
11. See Burnett 2012, 351.
12. On some of the benefits of recognition, see Markell 2003, 153. On this case in particular, see Deur 2000, 152.
13. An excerpt from *Makah Nation: A Whaling People*, a film by the Makah Whaling Commission. What Janine's friend apparently does not consider is that discourses on authenticity arise partly due to the nature of the claim: the Makah request for a cultural exemption invites scrutiny by its very form.
14. For a fuller discussion of the implicit view that indigenous politics are consensual rather than dialectical, and its effects, see Raibmon 2005, 12.
15. For a skeptical response to this question, see Hawkins (2001), who suggests that Makah claims about the cultural necessity of whaling represent a "selective labeling" and a retrenchment of cultural essentialism (289–92).
16. See Markell (2003) on Will Kymlicka's attempt to untangle this knot by drawing a distinction between the "existence" and the "character" of a culture, which "press[es] Kymlicka toward the uncomfortable conclusion that the normatively important work of culture is done by culture *qua* determining force: a right to one's own culture, it seems, depends on culture being just the sort of thing Herder and the anthropologists had thought" (160).
17. See Markell 2003, 175.
18. See, e.g., Coté 2010; Reid 2015.
19. See Jasanoff 2012.
20. A full-throated Marxian analysis might suggest that the exemption of whales from circulation offends the logic of capital that is the law's current patron.
21. The comments that NMFS selected, which they describe as a "representative sampling of 10 [sets of] comments that raise the most substantive issues" (NMFS 2015, 2) consisted of accusations of agency bias or corruption (14 percent), ecological concerns (16 percent), legal questions (11 percent), methodological critiques of the science used by NMFS in reaching its decision (18 percent), other procedural objections (21 percent), direct refutations of Makah public claims about whaling (6 percent),

concerns about the relationship between Makah whaling and commercial whaling (2 percent), animal welfare concerns (remarks about the suffering of hunted whales; 2 percent), other concerns about human impacts (e.g., economic impacts; 9 percent), and moral objections to killing whales (<1 percent). These figures are based on an original analysis using inductive coding of each individual "comment" as delineated by the NMFS report (each commenter in the sample submitted between 2 and 123 comments as defined by NMFS, with an average of 52).

BIBLIOGRAPHY

Abu-Lughod, Lila. 1991. "Writing against Culture." In Richard Fox, ed., *Recapturing Anthropology: Working in the Present*, 466–79. Santa Fe, NM: School of American Research Press.
Allen, B. M., and R. P. Angliss. 2011. "Stock Assessment Report. Gray Whale: Western North Pacific Stock." National Marine Fisheries Service, https://www.fisheries.noaa.gov/web dam/download/75294420.
Andersen, Peggy. 1998. "Elder Opposed to Whaling Finds Resistance at Home, Reverence Outside." *Los Angeles Times*, October 4, http://articles.latimes.com/1998/oct/04/local/me-29124.
Anderson, Virginia. 2004. *Creatures of Empire: How Domestic Animals Transformed America*. Oxford: Oxford University Press.
Anderson, Will. 2006. "Tribal Whaling Poses New Threat." In Nora Haenn and Richard Wilk, eds., *The Environment in Anthropology: A Reader in Ecology, Culture, and Sustainable Living*, 386–89. New York: New York University Press.
Arima, Eugene, and Alan Hoover. 2011. *The Whaling People of the West Coast of Vancouver Island and Cape Flattery*. Victoria: Royal BC Museum.
Barrett-Lennard, L. G., C. Matkin, D. Ellifrit, J. Durban, and L. Mazzuca. 2005. "Black and White versus Gray: Estimating Kill Rates, Consumption Rates, and Population-Level Impacts of Transient Killer Whales Feeding on Gray Whales." Abstract presented at the 16th Biennial Conference on the Biology of Marine Mammals, San Diego.
Barton, Karen Samantha. 2000. "'Red Waters': Contesting Marine Space as Indian Place in the United States Pacific Northwest." Master's thesis, University of Arizona.
Beck, Alma Soongi. 1996. "The Makah's Decision to Reinstate Whaling: When Conservationists Clash with Native Americans over an Ancient Hunting Tradition." *Journal of Environmental Law & Litigation* 11 (2): 359–412.
Bennett, Jane. 2009. *Vibrant Matter: A Political Ecology of Things*. Durham, NC: Duke University Press.
Berkes, Fikret, Johan Colding, and Carle Folke. 2000. "Rediscovery of Traditional Ecological Knowledge as Adaptive Management." *Ecological Applications* 10 (5): 1251–62.
Bessire, Lucas, and David Bond. 2014. "Ontological Anthropology and the Deferral of Critique." *American Ethnologist* 41 (3): 440–56.
Blok, Anders. 2007. "Actor-Networking Ceta-sociality, or, What Is Sociological about Contemporary Whales?" *Distinktion: Scandinavian Journal of Social Theory* 8 (2): 65–89.

Blore, Shawn. 1996. "Whaling Nation." *The Georgia Straight*, September 26. Available at http://www.shawnblore.com/Pieces/GeorgiaStraight/Makah/Makah1.htm.

Blow, Richard. 1998. "The Great American Whale Hunt." *Mother Jones* 23 (5): 49–53, 86–87.

Boas, Franz. 1897. *The Social Organization and the Secret Societies of the Kwakiutl Indians*. Washington: US National Museum. Available at http://archive.org/details/socialorganizat00huntgoog.

Bock, Paula. 1995. "The Accidental Whale—A Makah Fisherman Casts for a Long Submerged Treaty Right and Incidentally Nets a Whale." *Seattle Times*, November 26. Available at http://community.seattletimes.nwsource.com/archive/?date=19951126&slug=2154422.

Bowechop, Janine. 2004. "Contemporary Makah Whaling." In Marie Mauzé, Michael E. Harkin, and Sergei Kan, eds., *Coming to Shore: Northwest Coast Ethnology, Traditions, and Visions*, 407–20. Lincoln: University of Nebraska Press.

Brightman, Robert A. 1993. *Grateful Prey: Rock Cree Human-Animal Relationships*. Berkeley: University of California Press.

Brody, Hugh. 2000. *The Other Side of Eden: Hunters, Farmers, and the Shaping of the World*. New York: Macmillan.

Brown, Michael. 2009. *Who Owns Native Culture?* Cambridge, MA: Harvard University Press.

Burnett, D. Graham. 2012. *The Sounding of the Whale: Science and Cetaceans in the Twentieth Century*. Chicago: University of Chicago Press.

Busch, Briton Cooper. 1987. *The War against the Seals: A History of the North American Seal Fishery*. Montreal: McGill-Queen's University Press.

Butler, Judith. 1990. *Gender Trouble: Feminism and the Subversion of Identity*. New York: Routledge.

Callon, Michel. 1986. "Some Elements of a Sociology of Translation: Domestication of the Scallops and the Fishermen of St Brieuc Bay." In J. Law, ed., *Power, Action and Belief: A New Sociology of Knowledge?*, 196–223. London: Routledge.

———. 2010. "Egypt and the Experts." *Gerer & Comprendre* 100: 82–95.

Candea, Matei. 2010. "Ontology Is Just Another Word for Culture." *Critique of Anthropology* 30 (2): 172–79.

———. 2014. "The Ontology of the Political Turn." *Cultural Anthropology* online, January 13, https://culanth.org/fieldsights/469-the-ontology-of-the-political-turn.

Caras, Roger A. 1971. *Death as a Way of Life*. Boston: Little, Brown.

Cassaniti, Julia, and Jacob Hickman. 2014. "New Directions in the Anthropology of Morality." *Anthropological Theory* 14 (3): 251–62.

Cohn, Carol. 1987. "Sex and Death in the Rational World of Defense Intellectuals." *Signs* 12 (4): 687–718.

Collins, Cary C. 1996. "Subsistence and Survival: The Makah Indian Reservation, 1855–1933." *Pacific Northwest Quarterly* 87 (4): 180–93.

———. 2012. "The Water Is Our Land: The Diya Treaty Council of 1855." *Pacific Northwest Quarterly* 104 (1): 21–39.

Colson, Elizabeth. 1953. *The Makah Indians: A Study of an Indian Tribe in Modern American Society*. Minneapolis: University of Minnesota Press.

Comaroff, John L., and Jean Comaroff. 2009. *Ethnicity, Inc*. Chicago: University of Chicago Press.

Coté, Charlotte. 2002. "'It's Who We Are.' Makah and Nuu-chah-nulth Whaling: A History." PhD diss., University of California, Berkeley.

———. 2010. *Spirits of Our Whaling Ancestors: Revitalizing Makah and Nuu-chah-nulth Traditions*. Vancouver: University of British Columbia Press.

Coulthard, Glen. 2014. *Red Skin, White Masks: Rejecting the Colonial Politics of Recognition*. Minneapolis: University of Minnesota Press.

Coupland, Gary. 1998. "Maritime Adaptation and Evolution of the Developed Northwest Coast Pattern on the Central Northwest Coast." *Arctic Anthropology* 35 (1): 36–56.

Cronon, William. 1996. *Uncommon Ground: Rethinking the Human Place in Nature*. New York: W. W. Norton.

Curtis, Edward S. 1911. *The North American Indian: Being a Series of Volumes Picturing and Describing the Indians of the United States and Alaska*, vol. 11. Norwood, MA: Plimpton Press.

D'Amato, Anthony, and Sudhir Chopra. 2010. "Whales: Their Emerging Right to Life." *American Journal of International Law* 85 (10-32): 21–62.

D'Andrade, Roy. 1995. "Moral Models in Anthropology." *Current Anthropology* 36 (3): 399–408.

Darling, James. 1984. "Gray Whales off Vancouver Island, British Columbia." In M. L. Jones, S. L. Swartz, and S. Leatherwood, eds., *The Gray Whale*, 267–87. Orlando: Academic Press.

Daston, Lorraine. 1995. "The Moral Economy of Science." *Osiris* 10: 2–24.

Deloria, Vine. 1995. *Red Earth, White Lies: Native Americans and the Myth of Scientific Fact*. Golden, CO: Fulcrum.

Derrida, Jacques. 1991. "'Eating Well,' or the Calculation of the Subject: An Interview with Jacques Derrida." In Eduardo Cadava, Peter Connor, and Jean-Luc Nancy, eds., *Who Comes after the Subject?*, 97–119. New York: Routledge.

———. 2008. *The Animal That Therefore I Am*. New York: Fordham University Press.

Descola, Philippe. 2013. *Beyond Nature and Culture*. Chicago: University of Chicago Press.

Deur, Douglas. 2000. "The Hunt for Identity: On the Contested Targets of Makah Whaling." In Jim Norwine and Jonathan M. Smit, eds., *Worldview Flux: Perplexed Values among Postmodern Peoples*, 145–75. Lanham, MD: Lexington Books.

Diamond, Cora. 2004. "Eating Meat and Eating People." In Cass R. Sunstein and Martha C. Nussbaum, eds., *Animal Rights: Current Debates and New Directions*, 93–107. Oxford: Oxford University Press.

Doig, Ivan. 1982. *Winter Brothers: A Season at the Edge of America*. New York: Harcourt Brace Jovanovich.

Doniger, Wendy. 2014. *On Hinduism*. Oxford: Oxford University Press.

Dougherty, John. 2001. "Resurrection." *Phoenix New Times*, July 12, https://www.phoenixnewtimes.com/news/resurrection-6415044

Drucker, Philip. 1951. *The Northern and Central Nootkan Tribes*. Smithsonian Institution Bureau of American Ethnology, Bulletin 144. Washington: United States Government Printing Office.

Dupré, John. 1999. "Are Whales Fish?" In Douglas L. Medin and Scott Atran, eds., *Folkbiology*, 461–76. Cambridge, MA: MIT Press.

Einarsson, Niels. 1993. "All Animals Are Equal but Some Are Cetaceans: Conservation and Culture Conflict." In Kay Milton, ed., *Environmentalism: The View from Anthropology*, 73–84. London: Routledge.

Epstein, Charlotte. 2008. *The Power of Words in International Relations: Birth of an Antiwhaling Discourse*. Cambridge, MA: MIT Press.

Erikson, Patricia Pierce. 1999. "A-Whaling We Will Go: Encounters of Knowledge and

Memory at the Makah Cultural and Research Center." *Cultural Anthropology* 14 (4): 556–83.

———. 2002. *Voices of a Thousand People: The Makah Cultural and Research Center*. Lincoln: University of Nebraska Press.

Escobar, Arturo. 1999. "After Nature." *Current Anthropology* 40 (1): 1–30.

Fassin, Didier. 2008. "Beyond Good and Evil? Questioning the Anthropological Discomfort with Morals." *Anthropological Theory* 8 (4): 333–44.

———. 2009. "Moral Economies Revisited." *Annales, Histoire, Sciences Sociales* 64 (6): 1237–66.

———. 2012. *A Companion to Moral Anthropology*. New York: Wiley.

Fitzgerald, Deborah. 2003. *Every Farm a Factory: The Industrial Ideal in American Agriculture*. New Haven: Yale University Press.

Frasier, T. R., S. M. Koroscil, B. N. White, and J.D. Darling. 2011. "Assessment of Population Substructure in Relation to Summer Feeding Ground Use in the Eastern North Pacific Gray Whale." *Endangered Species Research* 14 (1): 39–48.

Freeman, Milton, Lyudmila Bogoslovskaya, Richard Caulfield, Ingmar Edege, and Igore Krupnik. 1998. *Inuit, Whaling, and Sustainability*. Lanham, MD: AltaMira Press.

Gibbs, George. 1877. "Tribes of Western Washington and Northwestern Oregon." *Contributions to North American Ethnology* 1 (2): 240.

Gilbert, Frederick F., and Donald G. Dodds. 1987. *The Philosophy and Practice of Wildlife Management*. Malabar, FL: Krieger.

Golde, Helen. 2012. "Notice of Intent to Terminate the Existing Draft Environmental Impact Statement and Prepare a New Environmental Impact Statement." 77 Fed. Reg. 98, 29967–69.

Gorman, Richard William, Jr. 2000. "Whales, Guns, and Money? How Commercial and Ideological Considerations Influenced the *Seattle Times* Portrayal of the Makah Whale Hunt." Master's thesis, University of Arizona.

Gosho, M. E. 1999. "Report of the NMFS Observer Monitoring the Makah Gray Whale Spring Hunt in 1999." Unpublished National Marine Fisheries Service–National Marine Mammal Laboratory report.

Gottlieb, Paul. 2012. "Makah Leader Disappointed by Whaling Study." *Peninsula Daily News*, May 25–26.

Greenwald, Glenn. 2017. "FBI Hunt for Two Missing Piglets Reveals the Federal Cover-Up of Barbaric Factory Farms." *The Intercept*, October 5, https://theintercept.com/2017/10/05/factory-farms-fbi-missing-piglets-animal-rights-glenn-greenwald/.

Gunther, Erna. 1936. "A Preliminary Report on the Zoological Knowledge of the Makah." In R. H. Lowie, ed., *Essays in Anthropology Presented to A.L. Kroeber in Celebration of His 60th Birthday, June 11, 1936*, 105–8. Berkeley: University of California Press.

———. 1942. "Reminiscences of a Whaler's Wife." *Pacific Northwest Quarterly* 3 (1): 65–69.

Gupta, Akhil. 2012. *Red Tape: Bureaucracy, Structural Violence, and Poverty in India*. Durham, NC: Duke University Press.

Hannula, Don. 1995. "The Whale Hunt: Listen to Case for the Makahs." *Seattle Times*, May 25. Available at http://community.seattletimes.nwsource.com/archive/?date=19950525&slug=2122946.

Harkin, Michael. 2007. "Swallowing Wealth: Northwest Coast Beliefs and Ecological Practices." In Michael Harkin and David Lewis, eds., *Native Americans and the Environment: Perspectives on the Ecological Indian*, 211–32. Lincoln: University of Nebraska Press.

Harkin, Michael, and David Lewis. 2007. *Native Americans and the Environment: Perspectives on the Ecological Indian*. Lincoln: University of Nebraska Press.

Harmon, Alexandra. 1998. *Indians in the Making: Ethnic Relations and Indian Identities around Puget Sound*. Berkeley: University of California Press.

Hawkins, Ronnie. 2001. "Cultural Whaling, Commodification, and Culture Change." *Environmental Ethics* 23 (3): 287–306.

Hays, Samuel P. 1959. *Conservation and the Gospel of Efficiency*. Cambridge, MA: Harvard University Press.

Heavey, Bill. 2013. "Hunting Your Own Dinner." *New York Times*, April 5, http://www.nytimes.com/2013/04/07/magazine/hunting-your-own-dinner.html.

Horwitz, Joshua. 2014. *War of the Whales*. New York: Simon & Schuster.

Howe, Nicolas. 2016. *Landscapes of the Secular: Law, Religion, and American Sacred Space*. Chicago: University of Chicago Press.

Huelsbeck, David R. 1988. "Whaling in the Precontact Economy of the Central Northwest Coast." *Arctic Anthropology* 25 (1): 1–15.

———. 1994. "Mammals and Fish in the Subsistence Economy of Ozette." In Stephan R. Samuels, ed., *Ozette Archaeological Reports*, vol. 2, 17–92. Pullman, WA: Washington State University Department of Anthropology.

Hughes, Johnson Donald. 1983. *American Indian Ecology*. El Paso: Western Press.

Hunn, Eugene, Darryl Johnson, Priscilla Russell, and Thomas Thornton. 2003. "Huna Tlingit Traditional Environmental Knowledge, Conservation, and the Management of a 'Wilderness' Park." *Current Anthropology* 44 (5): 79–103.

Ingold, Tim. 1987. *The Appropriation of Nature: Essays on Human Ecology and Social Relations*. Iowa City: University of Iowa Press.

———. 1989. "The Social and Environmental Relations of Human Beings and Other Animals." In Valerie Standen and Robert Foley, eds., *Comparative Socioecology: The Behavioral Ecology of Humans and Other Mammals*, 495–512. Oxford: Blackwell.

———. 2000. *The Perception of the Environment: Essays on Livelihood, Dwelling and Skill*. London: Routledge.

International Whaling Commission. 2012. *International Convention for the Regulation of Whaling, 1946: Schedule as Amended and Reorganized by the Commission at the 64th Annual Meeting, July 2012, and Replacing That Dated July 2011*. International Whaling Commission.

———. 2014. "History and Purpose." Accessed May 11, http://iwc.int/history-and-purpose.

Isaac, Barry L. 1988. "Introduction." In Barry L. Isaac, ed., *Research in Economic Anthropology, Supplement 3: Prehistoric Economies of the Pacific Northwest Coast*, 1–16. Greenwich, CT: JAI Press.

Jasanoff, Sheila. 1990. *The Fifth Branch: Science Advisers as Policymakers*. Cambridge, MA: Harvard University Press.

———. 2004. *States of Knowledge: The Co-production of Science and Social Order*. London: Routledge.

———. 2012. "Taking Life: Private Rights in Public Nature." In Kaushik Sunder Rajan, ed., *Lively Capital: Biotechnologies, Bioethics, and Governance in Global Markets*, 155–83. Durham, NC: Duke University Press.

Jewitt, John Rodgers, and Richard Alsop. 1815. *A Narrative of the Adventures and Sufferings, of John R. Jewitt: Only Survivor of the Crew of the Ship Boston, During a Captivity of Nearly Three Years among the Savages of Nootka Sound: With an Account of the Manners, Mode of Living, and Religious Opinions of the Natives*. Middletown, CT: Seth Richards.

Johnson, Keith. 1998. "The Makah Manifesto." *Seattle Times*, August 23. Available at http://community.seattletimes.nwsource.com/archive/?date=19980823&slug=2768010.

Kalland, Arne. 2009. *Unveiling the Whale: Discourses on Whales and Whaling*. New York: Berghahn Books.

Kalland, Arne, and Brian Moeran. 1992. *Japanese Whaling: End of an Era?* London: Curzon Press.

Kaste, Martin. 2014. "Tribe Views Whale Hunters as Heroes, Nuisances." NPR.org, March 12, http://www.npr.org/templates/story/story.php?storyId=88159568.

Kellert, Stephen R. 1979. *Public Attitudes toward Critical Wildlife and Natural Habitat Issues. Phase I*. Washington: US Fish and Wildlife Service.

———. 1980. "Contemporary Values of Wildlife in American Society." In W. W. Shaw and E. H. Zube, eds., *Wildlife Values*, 31–60. Tucson, AZ: Center for Assessment of Noncommodity Natural Resource Values.

Kim, Claire Jean. 2015. *Dangerous Crossings: Race, Species, and Nature in a Multicultural Age*. Cambridge, UK: Cambridge University Press.

Kirk, Ruth, and Richard D. Daugherty. 1974. *Hunters of the Whale: An Adventure of Northwest Coast Archaeology*. New York: Morrow.

Kirk, Ruth, and Meredith Parker. 2015. *Ozette: Excavating a Makah Whaling Village*. Seattle: University of Washington Press.

Kirksey, S. Eben, and Stefan Helmreich. 2010. "The Emergence of Multispecies Ethnography." *Cultural Anthropology* 25 (4): 545–76.

Kohn, Eduardo. 2013. *How Forests Think: Toward an Anthropology Beyond the Human*. Berkeley: University of California Press.

Kopytoff, Igor. 1986. "The Cultural Biography of Things: Commoditization as Process." In Arjun Appadurai, ed., *The Social Life of Things: Commodities in Cultural Perspective*, 64–91. Cambridge, UK: Cambridge University Press.

Kosek, Jake. 2006. *Understories: The Political Life of Forests in Northern New Mexico*. Durham, NC: Duke University Press.

Krech, Shepard. 2000. *The Ecological Indian: Myth and History*. New York: W. W. Norton.

Kulick, Don. 2003. "No." *Language and Communication* 23 (2): 139–51.

Laidlaw, James. 2001. "For an Anthropology of Ethics and Freedom." *Journal of the Royal Anthropological Institute* 8 (2): 311–32.

Landsman, Gail. 1988. *Sovereignty and Symbol: Indian-White Conflict at Ganienkeh*. Albuquerque: University of New Mexico Press.

Lang, A. R., B. L. Taylor, J. C. Calambokidis, V. L. Pease, A. Klimek, J. Scordino, K. M. Robertson, et al. 2011. "Assessment of Stock Structure among Gray Whales Utilizing Feeding Grounds in the Eastern North Pacific." Paper SC/M11/AWMP4 presented to the International Whaling Commission Scientific Committee.

Latour, Bruno. 1983. "Give Me a Laboratory and I Will Raise the World." In Karen D. Knorr-Cetina and Michael Mulkay, eds., *Science Observed: Perspectives on the Social Study of Science*, 141–70. London: Sage.

———. 1987. *Science in Action: How to Follow Scientists and Engineers through Society*. Cambridge, MA: Harvard University Press.

———. 2004. *Politics of Nature*. Cambridge, MA: Harvard University Press.

———. 2014. "Agency in the Time of the Anthropocene." *New Literary History* 45 (1): 1–18.

Lavigne, David, Victor Scheffer, and Stephen Kellert. 1999. "The Evolution of North American Attitudes toward Marine Mammals." In John Twiss and Randall Reeves, eds., *Conservation and Management of Marine Mammals*, 10–47. Washington: Smithsonian Institution Press.

Leopold, Aldo. 1933. *Game Management*. New York: Scribner.
Levy, Robert, and Douglas Hollan. 1998. "Person-Centered Interviewing and Observation in Anthropology." In H. R. Bernard, ed., *Handbook of Methods in Cultural Anthropology*. Lanham, MD: AltaMira Press.
Londoño-Sulkin, Carlos. 2012. *People of Substance*. Toronto: University of Toronto Press.
Makah Cultural and Research Center. 2010. *A Gift from the Past* (VHS). Neah Bay, WA: Makah Cultural and Research Center.
Makah Tribal Council. 2005. "Application for a Waiver of the Marine Mammal Protection Act Take Moratorium to Exercise Gray Whale Hunting Rights Secured in the Treaty of Neah Bay." Presented to the National Marine Fisheries Service.
Mapes, Linda V. 2002. "Makah Leaders Say More Pressing Needs than Whale Hunts Face Their People." *Seattle Times*, April 15. Available at http://community.seattletimes.nwsource.com/archive/?date=20020415&slug=makah15m.
———. 2007. "Makah Tribal Officials Dismayed over Whale Kill; Whaler Captain Has No Regrets." *Seattle Times*, September 9, https://www.seattletimes.com/seattle-news/makah-tribal-officials-dismayed-over-whale-kill-whaler-captain-has-no-regrets/.
Markell, Patchen. 2003. *Bound by Recognition*. Princeton, NJ: Princeton University Press.
Marker, M. 2006. "After the Makah Whale Hunt: Indigenous Knowledge and Limits to Multicultural Discourse." *Urban Education* 41 (5): 482–505.
Martello, Marybeth Long. 2002. "Negotiating Global Nature and Local Culture: The Case of Makah Whaling." In Sheila Jasanoff and Marybeth Long Martello, eds., *Earthly Politics: Local and Global in Environmental Governance*, 263–84. Cambridge, MA: MIT Press.
Marvin, Garry. 2010. "Wolves in Sheep's (and Others') Clothing." In Dorothee Brantz, ed., *Beastly Natures: Animals, Humans, and the Study of History*, 59–78. Charlottesville: University of Virginia Press.
Maté, B. R., V. Y. Ilyashenko, A. L. Bradford, V. V. Vertyankin, G. A. Tsidulko, V. V. Rozhnov, and L. M. Irvine. 2015. "Critically Endangered Western Gray Whales Migrate to the Eastern North Pacific." *Biology Letters* 11 (4): article no. 20150071.
McCarty, Micah. 2003. "Antiwhaling Agenda Ignores Nature's Realities." *Peninsula Daily News*, February 7.
McMillan, Alan Daniel. 1999. *Since the Time of the Transformers: The Ancient Heritage of the Nuu-chah-nulth, Ditidaht, and Makah*. Vancouver: University of British Columbia Press.
Menzies, Charles R. 1994. "Stories from Home: First Nations, Land Claims, and Euro-Canadians." *American Ethnologist* 21 (4): 776–91.
———. 2013. "Standing on the Shore with Saaban: An Anthropological Rapprochement with an Indigenous Intellectual Tradition." *Collaborative Anthropologies* 6 (1): 171–99.
Menzies, Charles, and Caroline Butler. 2006. "Understanding Ecological Knowledge." In C. R. Menzies, ed., *Traditional Ecological Knowledge*, 1–20. Lincoln: University of Nebraska Press.
Merton, Robert K. 1938. "Science and the Social Order." *Philosophy of Science* 5 (3): 321–37.
Miller, Robert J. 2000. "Exercising Cultural Self-Determination: The Makah Indian Tribe Goes Whaling." *American Indian Law Review* 25 (2): 165–273.
Mol, Annemarie. 2002. *The Body Multiple: Ontology in Medical Practice*. Durham, NC: Duke University Press.
Moore, Amanda Leslie. 2008. "Whale Stories: An Ethnography of Late Modern Nature." PhD diss., University of California, Irvine.
Mozino, Jose Mariano, and Iris H. Wilson Engstrand. 1991. *Noticias de Nutka*. Seattle: University of Washington Press.

Muehlmann, Shaylih. 2009. "How Do Real Indians Fish? Neoliberal Multiculturalism and Contested Indigeneities in the Colorado Delta." *American Anthropologist* 111 (4): 468–79.

Nadasdy, Paul. 2004. *Hunters and Bureaucrats: Power, Knowledge, and Aboriginal-State Relations in the Southwest Yukon.* Vancouver: University of British Columbia Press.

———. 2005. "Transcending the Debate over the Ecologically Noble Indian: Indigenous Peoples and Environmentalism." *Ethnohistory* 52 (2): 291–331.

———. 2007. "The Gift in the Animal: The Ontology of Hunting and Human-Animal Sociality." *American Ethnologist* 34 (1): 25–43.

National Marine Fisheries Service. 1997. "Environmental Assessment of the Makah Tribe's Harvest of up to Five Gray Whales per Year for Cultural and Subsistence Use."

——— 2013. "Stock Assessment Report. Gray Whale: Western North Pacific Stock." https://www.fisheries.noaa.gov/webdam/download/70078784.

———. 2014. "NOAA Fisheries Strategic Plan—Our Goals and Objectives." Accessed May 25, http://www.nmfs.noaa.gov/om2/goals.html.

———. 2015a. "Draft Environmental Impact Statement on the Makah Tribe Request to Hunt Gray Whales." US Department of Commerce.

———. 2015b. "Stock Assessment Report. Gray Whale: Western North Pacific Stock." https://www.fisheries.noaa.gov/webdam/download/70078785.

Nesper, Larry. 2002. *The Walleye War: The Struggle for Ojibwe Spearfishing and Treaty Rights.* Lincoln: University of Nebraska Press.

Nesper, Larry, and James Schlender. 2007. "The Politics of Cultural Revitalization and Intertribal Resource Management." In Michael Harkin and David Lewis, eds., *Native Americans and the Environment,* 277–303. Lincoln: University of Nebraska Press.

Nixon, Rob. 2011. *Slow Violence and the Environmentalism of the Poor.* Cambridge, MA: Harvard University Press.

Ollikainen, Rob. 2009. "Whale Killed in Illegal 2007 Hunt Identified." *Peninsula Daily News,* May 8, http://www.peninsuladailynews.com/news/whale-killed-in-illegal-2007-hunt-identified/.

Paine, R. T. 1969. "A Note on Trophic Complexity and Community Stability." *American Naturalist* 103 (929): 91–93.

Pearson, Susan, and Mary Weismantel. 2010. "Does 'the Animal' Exist? Toward a Theory of Social Life with Animals." In Dorothee Brantz, ed., *Beastly Natures: Animals, Humans, and the Study of History,* 17–37. Charlottesville: University of Virginia Press.

Petersen, Shannon. 1999. "Congress and Charismatic Megafauna: A Legislative History of the Endangered Species Act." *Environmental Law* 29 (2): 463–91.

Peterson, Brenda. 1996. "Who Will Speak for the Whales? Elders Call for a Spiritual Dialogue on Makah Tribe's Whaling Proposal." *Seattle Times,* December 22. Available at http://community.seattletimes.nwsource.com/archive/?date=19961222&slug=2366339.

Peterson, Brenda, and Linda Hogan. 2003. *Sightings: The Gray Whales' Mysterious Journey.* Washington: National Geographic Society.

Peterson, Melissa, and Makah Cultural and Research Center. 2003. "Makah." In Jacilee Wray, ed., *Native Peoples of the Olympic Peninsula: Who We Are,* 151–67. Norman: University of Oklahoma Press.

Peyton, R. B., and E. E. Langenau. 1985. "A Comparison of Attitudes Held by BLM Biologists and the General Public Toward Animals." *Wildlife Society Bulletin* 13 (2): 117–20.

Potter, Will. 2013. *Green Is the New Red: An Insider's Account of a Social Movement Under Siege.* San Francisco: City Lights Books.

Povinelli, Elizabeth. 1995. "Do Rocks Listen? The Cultural Politics of Apprehending Australian Aboriginal Labor." *American Ethnologist* 97 (3): 505–18.
Raibmon, Paige. 2005. *Authentic Indians*. Durham, NC: Duke University Press.
Ranco, Darren. 2007. "The Indian Ecologist and the Politics of Representation: Critiquing the Ecological Indian in the Age of Ecocide." In Michael Harkin and David Lewis, eds., *Perspectives on the Ecological Indian: Native Americans and the Environment*, 32–51. Lincoln: University of Nebraska Press.
Rappaport, Roy. 1968. *Pigs for the Ancestors: Ritual in the Ecology of a New Guinea People*. New Haven: Yale University Press.
Reid, Joshua. 2015. *The Sea Is My Country: The Maritime World of the Makahs, and Indigenous Borderlands People*. New Haven: Yale University Press.
Renker, Ann. 1997. "Whale Hunting and the Makah Tribe: A Needs Statement." International Whaling Commission, Exhibit 49AS5. Aberdeen, Scotland.
———. 2007. "Whale Hunting and the Makah Tribe: A Needs Statement." International Whaling Commission, Exhibit 59ASW9. Anchorage, AK.
Renker, Ann, and Erna Gunther. 1990. "Makah." In William C. Sturtevant, ed., *Handbook of North American Indians: Northwest Coast*, 422–30. Washington: Government Printing Office.
Ritvo, Harriet. 1998. *The Platypus and the Mermaid: And Other Figments of the Classifying Imagination*. Cambridge, MA: Harvard University Press.
Robertson, R. B. 1954. *Of Whales and Men*. New York: Knopf.
Ross, Debbie. 1992. "Makahs Want to Hunt Seals Again." *Peninsula Daily News*, November 13.
Russell, Dick. 2001. *Eye of the Whale: Epic Passage from Baja to Siberia*. Washington: Island Press.
Ryan, John. 2018. "An Ancient Fish Hook Could Solve a Big, Modern Problem." KUOW.org, April 26, http://archive.kuow.org/post/ancient-fish-hook-could-solve-big-modern-problem.
Sahlins, Marshall. 2013. *What Kinship Is—and Is Not*. Chicago: University of Chicago Press.
Said, Edward. 1978. *Orientalism*. New York: Pantheon Books.
Sapir, E. 1915. "The Social Organization of the West Coast Tribes." *Transactions of the Royal Society of Canada* 2 (9): 355–74.
Scarff, James E. 1980. "Ethical Issues in Whale and Small Cetacean Management." *Environmental Ethics* 3: 241–79.
Scheper-Hughes, Nancy. 1991. *Death without Weeping: The Violence of Everyday Life in Brazil*. Berkeley: University of California Press.
Schevill, William, G. Carleton Ray, and Kenneth S. Norris. 1974. *The Whale Problem: A Status Report*. Cambridge, MA: Harvard University Press.
Scordino, Jonathan J. 2007. "Report on Concerns Raised at September 11, 2007 Meeting on Unauthorized Whale Hunt." Memorandum to Donna Darm, Office of Protected Resources, National Marine Fisheries Service, November 15.
Scordino, Jonathan J., John Bickham, John Brandon, and Adrianne Akmajian. 2014. "What Is the PCFG? A Review of Available Information." Paper presented to International Whaling Commission Scientific Committee. Available at https://www.researchgate.net/publication/328615856_What_is_the_PCFG_A_review_of_available_information.
Scordino, Jonathan J., Jim Carretta, and Paul Cottrell. 2014. "Bycatch and Ship Strikes of Gray Whales in U.S. and Canadian Waters, 2008–2012." International Whaling Commission document no. SC/65b/BRG21, 1-19, https://swfsc.noaa.gov/publications/CR/2014/2014Scordino.pdf.

Scordino, Jonathan J., and Bruce Maté. 2011. "Bycatch and Ship Strikes of Gray Whales on US West Coast 1990–2010 and in British Columbia 1990–1995." *Report of the 2011 Aboriginal Whaling Management Procedure Workshop with a Focus on Eastern Gray Whales*, SC/63/Rep. 2, Annex C, 20–26.

Scott, James C. 1990. *Domination and the Arts of Resistance*. New Haven: Yale University Press.

———. 1998. *Seeing Like a State: How Certain Schemes to Improve the Human Condition Have Failed*. New Haven: Yale University Press.

Searle, John. 2006. "Social Ontology: Some Basic Principles." *Anthropological Theory* 6 (1): 12–29.

Sepez, Jennifer. 2001. "Political and Social Ecology of Contemporary Makah Subsistence Hunting, Fishing and Shellfish Collecting Practices." PhD diss., University of Washington.

Sepez Aradanas, Jennifer. 1998. "Aboriginal Whaling: Biological Diversity Meets Cultural Diversity." *Northwest Science* 72 (2): 142–45.

Sharp, Lesley A. 2013. *The Transplant Imaginary: Mechanical Hearts, Animal Parts, and Moral Thinking in Highly Experimental Science*. Berkeley: University of California Press.

Shorter, David. 2012. "Indian." *Frequencies: A Collaborative Genealogy of Spirituality*. http://frequencies.ssrc.org/2012/01/03/indian/.

Shweder, Richard. 1991. "Post-Nietzschean Anthropology: The Idea of Multiple Objective Worlds." Chap. 1 in *Thinking Through Cultures*. Cambridge, MA: Harvard University Press.

Siebert, Charles. 2009. "Watching Whales Watching Us." *New York Times Magazine*, July 12, http://www.nytimes.com/2009/07/12/magazine/12whales-t.html.

Simpson, Audra. 2014. *Mohawk Interruptus: Political Life Across the Borders of Settler States*. Durham, NC: Duke University Press.

Smithsonian National Zoo. 2014. "American Bison." https://nationalzoo.si.edu/animals/american-bison.

Sponsel, Leslie Elmer. 2012. *Spiritual Ecology: A Quiet Revolution*. Santa Barbara: ABC-CLIO.

Stacey, Natasha, Johanna Karam, Mark Meekan, Samuel Pickering, and Jonathan Ninef. 2012. "Prospects for Whale Shark Conservation in Eastern Indonesia through Bajo Traditional Ecological Knowledge and Community-Based Monitoring." *Conservation & Society* 10 (1): 63–75.

Stone, Christopher D. 2010. *Should Trees Have Standing? Law, Morality, and the Environment*. Oxford: Oxford University Press.

Stone, Steve. 2015. "Staff Responses to Comments on the 2008 Draft Environmental Impact Statement on the Makah Tribe's Request to Hunt Gray Whales." National Marine Fisheries Service memorandum, February 19.

Sullivan, Robert. 1998. "Permission Granted to Kill a Whale. Now What?" *New York Times Magazine*, August 9, https://www.nytimes.com/1998/08/09/magazine/permission-granted-to-kill-a-whale-now-what.html.

———. 2000. *A Whale Hunt: How a Native-American Village Did What No One Thought It Could*. New York: Simon & Schuster.

Swan, James G. 1870. *The Indians of Cape Flattery: At the Entrance to the Strait of Fuca, Washington Territory*. Washington: Smithsonian Institution.

Tanner, Adrian. 1979. *Bringing Home Animals: Religious Ideology and Mode of Production of the Mistassini Cree Hunters*. St. John's: Institute of Social and Economic Research, Memorial University of Newfoundland.

Telesca, Jennifer. 2017. "Accounting for Loss in Fish Stocks: A Word on Life as a Biological Asset." *Environment and Society* 8: 144–60.
Thornton, Thomas, and Jamie Hebert. 2015. "Neoliberal and Neo-communal Herring Fisheries in Southeast Alaska: Reframing Sustainability in Marine Ecosystems." *Marine Policy* 61: 366–75.
Trouillot, Michel-Rolph. 2003. "Anthropology and the Savage Slot: The Poetics and Politics of Otherness." Chap. 1 in *Global Transformations: Anthropology and the Modern World*. New York: Palgrave Macmillan.
Tuan, Yi-Fu. 1984. *Dominance and Affection: The Making of Pets*. New Haven: Yale University Press.
Turner, Dale. 2006. *This Is Not a Peace Pipe: Towards a Critical Indigenous Philosophy*. Toronto: University of Toronto Press.
Van Ginkel, Rob. 2004. "The Makah Whale Hunt and Leviathan's Death." *Etnofoor* 17 (1/2) 58–89.
Verhovek, Sam H. 1999. "After the Hunt, Bitter Protest and Salty Blubber." *New York Times*, May 19, https://www.nytimes.com/1999/05/19/us/after-the-hunt-bitter-protest-and-salty-blubber.html.
Waterman, Thomas Talbot. 1920. *The Whaling Equipment of the Makah Indians*. Seattle: University of Washington Press.
Webb, Robert Lloyd. 1988. *On the Northwest: Commercial Whaling in the Pacific Northwest, 1790–1967*. Vancouver: University of British Columbia Press.
Weismantel, Mary. 2015. "Seeing Like an Archaeologist: Viveiros de Castro at Chavin de Huantar." *Journal of Social Archaeology* 15 (2): 139–59.
Weller, David W., Shannon Bettridge, Robert L. Brownell Jr., Jeffrey L. Laake, Jeffrey E. Moore, Patricia E. Rosel, Barbara L. Taylor, and Paul R. Wade. 2013. "Report of the National Marine Fisheries Service Gray Whale Stock Identification Workshop." NOAA technical memorandum, https://swfsc.noaa.gov/publications/TM/SWFSC/NOAA-TM-NMFS-SWFSC-507.pdf.
Wessen, Gary. 1982. "Shell Middens as Cultural Deposits: A Case Study from Ozette." PhD. diss., Washington State University.
Westneat, Danny. 1996. "Makah Closing Ranks around Whaling." *Seattle Times*, October 13. http://community.seattletimes.nwsource.com/archive/?date=19961013&slug=2354125.
Westneat, Danny. 2015. "Whale-Hunt Scolds Are Off Target." *Seattle Times*, March 13, https://www.seattletimes.com/seattle-news/whale-hunt-scolds-are-off-target/.
Whitehead, Hal, and Luke Rendell. 2014. *The Cultural Lives of Whales and Dolphins*. Chicago: University of Chicago Press.
Willerslev, Rane. 2007. *Soul Hunters: Hunting, Animism, and Personhood among the Siberian Yukaghirs*. Berkeley: University of California Press.
Wilson, Bryan R. 1991. *Rationality*. Oxford: Wiley-Blackwell.
Wohlforth, Charles. 2005. *The Whale and the Supercomputer: On the Northern Front of Climate Change*. New York: Farrar, Straus & Giroux.
Working Group on Whale Killing Methods and Associated Welfare Issues. 2012. "Report of the Working Group on Whale Killing Methods and Associated Welfare Issues." Presented at 64th Annual Meeting of the International Whaling Commission, Panama City, https://iwc.int/private/downloads/S-YJxeIrM5ar6t8ilp4RyA/64-Rep6.pdf.
Wynne, Brian. 2013. *Rationality and Ritual: Participation and Exclusion in Nuclear Decision-Making*. Abingdon, UK: Routledge.

Young, Donna J. 2005. "Writing against the Native Point of View." In Anne Meneley and Donna J. Young, eds. *Auto-Ethnographies: The Anthropology of Academic Practices*, 1–22 Peterborough, ON: Broadview Press.

Ziontz, Alvin J. 2009. *A Lawyer in Indian Country: A Memoir*. Seattle: University of Washington Press.

INDEX

Page numbers in italics denote figures.

animals, nonhuman: agency of, 8, 21, 167–68, 212n11; *becoming*, concept of, 3, 7, 191n8; as ethnographic subjects, 2–3, 7–8, 166–67, 191n7; and law, 182; as sacrificial structures, 183. *See also* antiwhaling activism; gray whales; killing (animals); Makah whaling; whales; whaling.

antiwhaling activism, 51, 63: adaptation to fisheries bureaucracy, 6, 136, 138–39, 144–51, 171–73, 180; adoption of gray whales, 13, 147, *147*; and aesthetics, 127, 179, 210n15; direct intervention in Makah whale hunt, 11, 12, 132–36, *134*; and ecocentrism, 133; history of, 125–27, 179; lawsuits, 27, 146, 212–13n5; motivations, 127–29, 133, 135, 179; Peninsula Citizens for the Protection of Whales, 27, 121–22, 129–32, *147*; and racism, 19, 123, 130, 193n35, 207n2; roadside protests, 11, 130–31, 135; slippery slope argument, 51, 87, 203n5; speaking for whales, 130, 168, 174–75, 183; stereotypes of, 123; stumbling blocks (antiwhaling tactic), 145–51; superwhale, 208n15; tactics, 124, 129–36, 145–51. *See also under* state power

Blok, Anders, 174, 178, 212n11
Boldt decision, 10, 43–44, 53, 87, 111, 200n89, 203n4, 205n2
Bowechop, Janine, 42, 81, 119, 157, 181–82

Brown, Michael, 22
bureaucracy. *See under* antiwhaling activism; Makah whaling. *See also* International Whaling Commission (IWC); management, fisheries and wildlife; National Marine Fisheries Service (NMFS); state power: technocratic; technocracy
Burnett, Graham, 17, 126, 142, 207n6, 207n9, 211n19

Calambokidis, John, 13. *See also* Cascadia Research Collective
Callon, Michel, 174
Candea, Matei, 82, 192n9
Cascadia Research Collective, 13, 147, 171, 208n1
ceremonial and subsistence (C&S) harvest, 27, 51, 86–87, 200n90, 203n4, 205n2
cetology, "hip-booted," 142
Coast Guard, US, 98, 172: arrest of rogue hunt whalers, 12, 155; moving exclusionary zone, 11, 107, 119, 135; seizure of antiwhaling activist vessels, 11, 119, 135–36, 180; support for Makah Tribe, 119, 207n24
Cohn, Carol, 144
Colfax, Greg, 45–46, 64–65, 88, 104–6
Colson, Elizabeth, 36–37, 41–42, 91, 198n55, 203n10, 205n12
Comaroff, Jean, 34, 193n33
Comaroff, John, 34, 193n33

Coté, Charlotte, 84, 87, 158, 204n14
Coulthard, Glen, 16, 109, 177, 180, 195n5, 206n19
CR-175 (whale), 13, 171
Cronon, William, 3, 123
culinary imperialism, 84
cultural whaling. *See* Makah whaling: cultural need, concept of
culture: as basis for exemption to law or regulation, 20, 181–82, 194n38, 197n40, 213n13, 213n16; IWC's concept of, 11, 181, 193n25; nature-culture binary, 2, 174

delisting of Eastern North Pacific gray whale, 10, 47–49, 199nn71–72
Derrida, Jacques, 183, 191n7, 192n11
Descola, Phillipe, 192n15
Diamond, Cora, 62
Dupré, John, 192n13

Einarsson, Niels, 207n1
Endangered Species Act, 48, 116–17, 126, 139
erasure by interpretation, 166
Erikson, Patricia, 24

fish wars. *See* Makah struggle for fishing rights
fisheries. *See* management, fisheries and wildlife

grateful prey. *See* spirituality, Makah: animals offering selves to hunters
gray whales, *143, 147*: approach, penchant to, 1–2, 153, 158, 162–65, 168, 171, 212n2 (*see also* gray whales: Friendlies); behavior, 152, 167 (*see also* gray whales: approach, penchant to); birthing lagoons, 70, 129, 153, 162–64, 183–84; breath of, 127, 152, 179, 207n12; charisma (lack of), 152; Friendlies, 2, 70, 153; 162–65, 171, 184, 212n2; historical exploitation by Makah, 47, 153, 192n18, 196n16, 212n3; meat, vs. humpback whale, 47, 61, 153; migration route, 13; psychology, 166–68; resident (*see* Pacific Coast Feeding Group); Western North Pacific, 184, 210n17
Gunther, Erna, 106, 202n12

Harkin, Michael, 200n87, 202n18
Harmon, Alexandra, 40–41, 197nn48–50
Hawkins, Ronnie, 213n15
Herschel (sea lion), 66
Hogan, Linda, 71,171, 207–8n14
Horwitz, Joshua, 162–63
Howe, Nicolas, 203n1
Huelsbeck, David, 196n16, 212n3

identity, 34–35, 59–60, 181, 195n8. *See also* Makah identity; Makah whaling: and identity
Ingold, Timothy, 89, 99–103, 203–4n13
International Whaling Commission (IWC), 50, 98, 118, 120, 125, 141–42, 172, 184: aboriginal subsistence whaling exemption, 10, 50, 181; Makah needs statement, 50, 201n108; Makah representation at, 58; meetings in Aberdeen in 1996, 10–11, 50, 69; meetings in Monaco in 1997, 51; moratorium on commercial whaling, 10, 17, 140–41; purpose and mission, 141–42; Scientific Committee, 118, 141, 174, 209n7
Isaac, Barry L., 196n11, 196n16, 197n46

Jasanoff, Sheila, 2, 140
Johnson, Wayne (Makah whaling captain), 11, 24, 34, *46*, 73, 92–93, 132, 153–57, 161

killing (animals): ambivalence about, 72–73; vs. asking animal for its life, 177–81; drawing lines for, 128; euphemisms for, 144.
Kim, Claire Jean, 20
Kohn, Eduardo, 192n15
Kopytoff, Igor, 178, 205n2, 213n9

Latour, Bruno, 3, 167–68, 174, 191n4

Makah community, 8, 18: animals, general attitudes toward, 66, 72, 178, 206n19; "culture," 25, 194n47; extended families as social unit, 18, 36, 40, 56–58, 64, 88–89, 198n51, 198n55; fisheries management discourse, 109–10; whaling families, 33–34, 58, 59–60, 98, 100, 157, 201n108
Makah Cultural and Research Center

(MCRC), 25, 27, 32, 56, 60, 104, 181, 199n65, 199n68
Makah Fisheries. *See* Makah resource management.
Makah identity, 32–36, 52, 70, 99, 176
Makah Reservation, 24–25
Makah resource management: affinity with Makah traditions, discourse on, 108, 111–17, 179; experience of commercial fishermen, 111; Makah as original ecological stewards, 111–13, 115–16; moral consonance with modern bureaucratic fisheries management, 117–18, 179; "take only what you need," 109, 201n10
Makah struggle for fishing rights 10, 42–45, 70, 77, 111, 176–77
Makah Tribe: history of, 8, 40; treated as monolithic, 18–19, 193–94n36
Makah whaling: and bureaucracy, 86–87, 98, 107, 116, 172–73; as distinction, 32, 42; commercial, 40, 47–48, 86–87, 99, 203n3; cultural need, concept of, 50–51, 76, 200n86; economic importance prior to contact, 36–37; elders opposed, 10, 50, 68–71, 69; family hunts (2000), 58, 153; in food economy prior to contact, 37, 196n16; as hereditary privilege, 34, 56, 58, 59, 157, 201nn107–8; history of, 8–9, 14, 36–42, 192n18, 193n20, 197n37, 197n42; hunt of 1999, 1, 12, 75, 99, 133, 200–201n101; and identity, 32–60; 176 (*see also* "Yum!": politics of); impetus for revival, 10, 45, 49, 50, 65, 176; importance of, 9, 10, 18, 38, 50; keystone species argument, 95–96, 110, 205n11; MMPA waiver request, 146; needs statement (IWC), 50; nutritional need, 50, 201n108; opposition as catalyst, 53–55, 59; oral histories, 42, 47, 90, 103, 104; as political act, 69, 162, 176; pressure to present unified view, 11, 19–20, 60, 71, 76; as "reaffirmation" of identity, 19, 28, 32, 35, 175–76, 194n1; and religious freedom, 87, 161, 176; rights in treaty, 12, 38, 52, 107, 153, 161, 172; rogue hunt, 1, 12, 13, 53, 77, 98, 153–57, 160–62, 172, 176, 193n29; science and, 48, 118, 172, 206n22; and social hierarchy, 8, 36, 55–56, 71, 192n19;
and spirituality, 65, 76, 77, 83, 85–106 (*see also* spirituality, Makah); spiritual preparation for, 69, 86, 87–88, 99, 102–3, 155, 159, 160–61, 203n2, 204n18; struggle for, importance of 35, 52–55, 175; subsistence need, 10, 50; support within tribe, 194–95n4; technology, 51, 96–99, 181; as tribal venture, 56, 200–1n101; voluntary cessation of, 9, 40; whaling families, 33–34, 58, 59–60, 98, 100, 157, 201n108; Yellow Banks whale, 62, 158
Makah whaling conflict, summary of, 8–13
management, fisheries and wildlife: use of agronomic metaphors, 143–44, 209n11; conservation, proper unit of, 140, 172–73, 210–11n18; "single-species," discourses on, 117; stock, concept of, 140; weak-stock, 138, 209n6
Marine Mammal Protection Act (MMPA), 9, 12, 45, 116–17, 126, 132, 135, 139, 140–41, 146–49, 172, 183
Markell, Patchen, 14–16, 59, 108, 181, 183, 194n38, 231n16
McCarty, Micah, 87, 90, 95–97, 104, 110–11, 119, 138, 202n14, 209n5
Miller, Robert, 32, 89
Mol, Annemarie, 191n5
morality as anthropological category, 6–7, 17, 21, 82, 178

Nadasdy, Paul, 81–82, 94, 108–9, 160, 178, 202n18
National Environmental Policy Act (NEPA), 142, 212–13n5: NMFS violation of, 12, 146
National Marine Fisheries Service (NMFS), 13, 182, 183–85: ecosystem-based management, 117, 206n20; environmental assessment of Makah whaling, 12, 14, 146; draft environmental impact statement on Makah whaling, 12, 13, 138, 144, 165, 184–85, 212–13n5; Makah whaling, oversight of, 10, 12, 13–17, 48, 50, 86–87, 107–8, *134*, 139, 142, 205n2; Makah whaling, support for, 49, 200n80; moral economy of, 14, 17, 28, 139–45, 150–51, 171–73, 179–80, 185, 206n21, 209–10n13, 213n21; National Environmental Policy Act, interpretation of, 142, 212–

National Marine Fisheries Service (NMFS) (*continued*)
13n5; and potential biological removal (PBR), 17, 143, 193n31; purpose and mission of, 140; quantifying impulse, 5, 17, 142–44; as technocracy, 140
Native American: identities, 10; tribal affiliations on the Northwest Coast, 9, 40–42
nature: limits to social constructivism, 3, 166–67, 191n7; social construction of, 2, 191nn4–5
nature-culture binary, 2, 174
NEPA. *See* National Environmental Policy Act
Nixon, Rob, 5
noble ecological Indian, myth of, 63, 81, 96, 113, 115, 202n15, 204n14, 206n16
NMFS. *See* National Marine Fisheries Service.
nonhuman persons, concept of, 82–83, 95–96, 178, 202n17, 203–4n13
Northwest Indian Fish Commission (NWIFC), 48

ontology: vs. epistemic approach, 3; ontological turn, 6–7, 93–94, 192n9, 192n16; as politics, 4, 82, 178
oral history, 90, 104. *See also under* Makah whaling
Owens, Charles "Chuck," 26–27, 121–24, *124*, 127–31, *128*, 133–35, 145–49, 207n13, 208n16
Owens, Margaret, 26–27, 70, 121–24, 127–36, *124*, 137–38, 144–50, 171, 179
Ozette archeological discovery, 44, 198n64: as catalyst for cultural revival, 44–45, 176, 199n66

Pacific Coast Feeding Group (PCFG), 138, 140, 146–50, *148*, 171–73, 184, 212–13n5: existence of, 3, 13–14, 146–47; genetic findings on, 13, 127–38, 147–49, 208n4, 210–11n18; naming of, 13, 137, 209n5; range/definition of, 13–14, 15, 137, 208n3; as stock or separate management unit, 13, 147–49, 209n6, 210n16
Parker, Theron, 99–104, 153–61, 168
Paul, Jonathan, 122, 132–36, 168–69
Pearson, Susan, 6, 212n10

personhood, nonhuman. *See* nonhuman persons, concept of
Peterson, Brenda, 171, 194–95n4, 199n72, 201n8, 207–8n14
politics. *See* Makah whaling: as political act, 69, 162, 176
politics of recognition. *See* recognition, politics of
potential biological removal (PBR), 17, 143, 193n31
Povinelli, Elizabeth, 109
preservationism, vs. conservation, 4, 17–18, 70, 72, 116–17, 140–41, 150, 163, 172–73, 182–83, 206–7n23, 211n20

recognition, politics of, 180–83, 197n40, 213n12
Reid, Joshua, 19, 113, 197n37
religious freedom and Makah whaling, 87, 161, 176
Renker, Ann, 50, 71, 87, 111, 201n108
resident gray whales. *See* Pacific Coast Feeding Group
Ritvo, Harriet, 2
rogue hunt. *See under* Makah whaling

"save the whales," 5, 202n14
science, fisheries, 137–39. *See* National Marine Fisheries Service (NMFS); Pacific Coast Feeding Group: genetic findings on; Makah whaling: science and. *See also* antiwhaling activism: lawsuits
scientific uncertainty, 125, 150, 172, 180, 211n19
Scott, James, 173
Sea Shepherd Conservation Society, 11, 51, 70, 122–24, *134*, 184, 207n3
sealing, Makah: contemporary, 45–46, 49, 78–79; history of, 39–40
Scordino, Jonathan, 149, 193n29, 199n67
Searle, John, 21
Sepez, Jennifer, 64
Shweder, Richard, 192n9
Siebert, Charles, 153, 162
Simpson, Audra, 35, 109, 180, 182, 194n49, 195n8
slave ancestry, discourse on, 58, 71
species boundary, 166
spirituality, Makah; animals offering selves to hunters, 1–2, 29, 74, 80–81, 93–96, 156–60, 191n1, 202n19; as

care, 96; and Christianity, 88, 90–92, 203n11; discourses on, 86, 178–79; and environmental stewardship, 92, 96, 97, 113–16; ethnographic sources, use of, 104; "everything is connected," 86, 89, 105–6; food and, 64–65, 103; and law, 203n1; material world, relation to 64–65, 96; and modern management, 113–16; as a problematic term, 86; as relationship with animal, 88, 93–97, 159–60; and sexual abstinence, 94, 102; as "skill," 89, 99–103; songs, 94, 102, 156; sweat lodges, 102, 103; and technology, 96–99; variability of, 87–89, 161; and wives of Makah whalers 105–6

state power, 12, 14–17, 116, 124–25, 177, 185: antiwhaling activism, effect on, 138–39, 144–51, 171–73; barriers to entry, 16–17, 136, 144–45; and indigenous rights, 177, 180–82; as technocratic, 17, 140; traditional ecological knowledge, as incorporating, 108–9, 177; as transformative, vs. additive, 144; use of legitimate violence, 12, 16, 135–36

Stone, Christopher, 210n15

strategic essentialism, 11, 19–20, 51, 60 195n5

Sullivan, Robert, 11, 93, 101, 107, 131–32, 155, 202n14

superwhale. *See under* antiwhaling activism

Swan, James, 36, 39, 104, 195n9, 197n37, 202n11

technocracy, 16–17, 108, 136, 140, 180, 210n14

Thompson, Alberta "Binki," 70–71, 76, 84, *163*

Thunderbird (and the whale), 32, 76, *76*, 83, 194n2, 202n11

traditional ecological knowledge, 108, 205n5, 205n7: and state power, 108–9, 177

translation, 6, 17, 108, 117–20, 137–51, 171–75, 183

Treaty of Neah Bay, 1855, 6, 8–9, 38, 177, 196nn30–31

treaty rights, discourse on 6, 35, 41, 53–55, 176–77, 184, 195n7, 200n100, 204–5n1

Tribal Canoe Journeys, 100

Turner, Dale, 177

United States v. Washington, 1974. *See* Boldt decision

usual and accustomed fishing and hunting grounds (U&A), Makah, 9, 10, 43

Van Ginkel, Rob, 20

Washington, state of: vilification of Makah as poachers, 42, 63, 77

Waterman, T. T., 37, 104, 197n37, 203n2

Watson, Paul. *See* Sea Shepherd Conservation Society

Weismantel, Mary, 6, 178, 212n10

whale watching, 165–66

whales: as charismatic megafauna, 9; conservation of, historical, 125–26; as edible, things-to-be-eaten, 73, 178 (*see also* "Yum!": ethics of); as fish, 5, 17, 144, 180, 192n13; as inscrutable surface upon which human interests are inscribed, 166; as killable, 5, 142–43, 151, 179, 183 (*see also* whales: as stocks); legal status of, 9; "on the beach," 22, 29, 73–76, *75*, 84, 152; as resources, 140–44, 150; popular opinion of, 9–10, 125–26, 175, 179–80, 192n13, 193n23, 207n6; as special, 127–28, 179; as stocks, 140–45, *143*, 150; as symbol of US environmental movement, 126, 140; as wild, 165–66

whaling, 74: commercial, 141, 213n6; Japanese, 18, 141–42, 203n5, 213n6; Norwegian, 18, 203n5; public opposition to, 9–10, 17–18, 119, 125–27, 175, 179–80

Willerslev, Rane, 202n17

Yabis (whale), 2–4, 191n6

"Yum!," 29, 61, *62*, 63–65, 84: ethics of, 63, 68–70, 178; politics of, 66–68, 72, 177–78

Ziontz, Al, 43, 51, 55